T0134563

Springer Theses

Recognizing Outstanding Ph.D. Research

Aims and Scope

The series "Springer Theses" brings together a selection of the very best Ph.D. theses from around the world and across the physical sciences. Nominated and endorsed by two recognized specialists, each published volume has been selected for its scientific excellence and the high impact of its contents for the pertinent field of research. For greater accessibility to non-specialists, the published versions include an extended introduction, as well as a foreword by the student's supervisor explaining the special relevance of the work for the field. As a whole, the series will provide a valuable resource both for newcomers to the research fields described, and for other scientists seeking detailed background information on special questions. Finally, it provides an accredited documentation of the valuable contributions made by today's younger generation of scientists.

Theses are accepted into the series by invited nomination only and must fulfill all of the following criteria

- They must be written in good English.
- The topic should fall within the confines of Chemistry, Physics, Earth Sciences, Engineering and related interdisciplinary fields such as Materials, Nanoscience, Chemical Engineering, Complex Systems and Biophysics.
- The work reported in the thesis must represent a significant scientific advance.
- If the thesis includes previously published material, permission to reproduce this must be gained from the respective copyright holder.
- They must have been examined and passed during the 12 months prior to nomination.
- Each thesis should include a foreword by the supervisor outlining the significance of its content.
- The theses should have a clearly defined structure including an introduction accessible to scientists not expert in that particular field.

More information about this series at http://www.springer.com/series/8790

Raphael Enoque Ferraz de Paiva

Gold(I,III) Complexes Designed for Selective Targeting and Inhibition of Zinc Finger Proteins

Doctoral Thesis accepted by the University of Campinas, Campinas, Brazil

Author
Dr. Raphael Enoque Ferraz de Paiva
Institute of Chemistry
University of Campinas
Campinas, Brazil

Supervisor
Prof. Pedro Paulo Corbi
Institute of Chemistry
University of Campinas
Campinas, Brazil

ISSN 2190-5053 ISSN 2190-5061 (electronic)
Springer Theses
ISBN 978-3-030-13150-0 ISBN 978-3-030-00853-6 (eBook)
https://doi.org/10.1007/978-3-030-00853-6

This Springer imprint is published by the registered company Springer Nature Switzerland AG
The registered company address is: Gewerbestrasse 11, 6330 Cham, Switzerland

The most exciting phrase to hear in science, the one that heralds the most discoveries, is not "Eureka!" (I found it!) but "That's funny..."

—Isaac Asimov

The mediocre teacher tells. The good teacher explains. The superior teacher demonstrates. The great teacher inspires

—William Arthur Ward

I dedicate this thesis to every teacher (in the most fundamental sense of the word) I had during my life. Luckily, I've had many great teachers.

Supervisor's Foreword

Raphael joined my group as an undergraduate student at the Institute of Chemistry of the Campinas State University (UNICAMP), Brazil, in 2010. He obtained his master's degree working in the synthesis and biological evaluation of silver and platinum complexes with anti-inflammatory agents. Raphael developed his Ph.D. project focusing on the development of gold(I,III) compounds for the inhibition of *zinc finger* proteins. This project was developed at UNICAMP with a one year exchange at VCU under the supervision of Prof. Nicholas Farrell.

The zinc finger family of proteins was explored as target in this work, in particular, the nucleocapsid protein (NCp7) from the human immunodeficiency virus (HIV-1). One of the most remarkable structural and functional characteristics NCp7 is the presence of two -Cys-X2-Cys-X4-His-X4-Cys- zinc finger (ZnF) domains. Inhibition of this proteins leads to the loss of viral infectivity, suggesting that NCp7 inhibitors could be developed as alternatives to the typical anti-retroviral therapies available today.

The broad contribution of this thesis was on the development of gold-based compounds for the inhibition of zinc finger proteins. Gold(I) compounds were developed using phosphine ligands as "carriers" and labile pyridine derivatives or chloride as co-ligands. Innovative approaches were used for evaluating the interaction of gold(I) compounds with zinc fingers. Travelling-wave ion mobility mass spectrometry (TWIM-MS) coupled with tandem MS proved to be an extremely powerful technique for detecting metallation sites in zinc fingers, as well as for identifying and separating conformers. X-ray absorption spectroscopy (XAS) was used for identifying changes in the coordination sphere and geometry of gold. In addition, gold(III) compounds were also developed. Stabilizing gold(III) represents an interesting challenge, as it is prone to reduction in the biological media. The organometallic compounds [Au(2-benzylpyridine)Cl_2] is particularly noteworthy. It was able to inhibit NCp7 through a *sui generis* mechanism based on a gold-catalyzed C-S bond formation.

Finally, the dedication and motivation of Dr. Raphael de Paiva are noteworthy. He was always willing to help his colleagues and ready for new challenges. His published works and conferences awards confirm his wonderful work and his dedication to science. It was a great honor to me to write about Dr. Raphael Enoque Ferraz de Paiva and I wish him success in his career.

Campinas, Brazil
July 2018

Prof. Pedro Paulo Corbi

Abstract

Many complementary approaches have been used to inactivate the HIV-1 nucleocapsid protein (NCp7) zinc finger for therapeutic use. The cysteine residues of NCp7 are some of the most nucleophilic of all zinc-bound thiolates found in proteins. Although it is considered a noble metal, gold has a very rich chemistry under the right conditions. It is capable of forming coordination compounds in both of its typical oxidation states (1+ and 3+). Gold(I) forms linear compounds, with coordination number 2, while gold(III) forms square planar compounds with coordination number 4. Electron-rich metals such as gold are particularly suitable for designing metal-based zinc ejectors because, as soft Lewis acid electrophiles, they have high affinity for Cys residues. Without the proper tridimensional folding granted by Zn coordination, NCp7 is inactivated and unable to further recognize specific nucleic acid sequences.

A series of Au(I) complexes with the general structure [Au(L)(PR_3)] (R = Et or Cy) was designed. The two aromatic residues in the structure of NCp7 (Phe16 and Trp37) are responsible for π-stacking with purine and pyrimidine bases found on RNA and DNA. The presence of these residues can be used for introducing an extra selectivity component in the designed compounds, as an aromatic L ligand can be used. We also examined for comparison the "standard" gold-phosphine compound auranofin which contains a thiosugar ligand coordinated to {Au(Et_3P)}. The nature of the phosphine and the nature of L affect both the reactivity with the C-terminal NCp7 ZnF2 and the "full" NCp7 as well as the final coordination sphere of Au once incorporated into the protein. In this reaction, the first step is the electrophilic attack of the Au(I)-phosphine compounds on the Zn-coordinated residues, forming a heterobimetallic {R_3PAu}-ZnF species. Two alternative pathways open up from here on. In the most typical pathway, after Zn displacement, the {R_3PAu} moiety can remain coordinated to a Cys residue. Afterwards, the phosphine is lost and a Cys-Au-Cys *gold finger* (AuF) is obtained. In this mechanism, compounds with slower reactivity such as [Au(dmap)(Et_3P)] allow us to probe the initial auration steps, with species such as {Et_3PAu}-apoNCp7 still being present. Longer incubation times or more reactive compounds such as the chloride precursors provide information on the final species, the gold finger itself. In an alternative pathway, observed for the model compound auranofin, the {R_3PAu} moiety coordinates to a His residue, and the final hypothetical AuF obtained from this pathway has Cys-Au-His coordination sphere. The Au(I)-phosphine series also had its cytotoxic properties investigated. The compound $[Au(dmap)(Et_3P)]^+$ demonstrated a cytotoxic selectivity >50 towards a tumorigenic T lymphoblast cell line (CEM) in comparison to a normal cell line (HUVEC). The compound $[Au(dmap)(Et_3P)]^+$ caused apoptosis on the CEM cell line through a mechanism independent of p38 MAPK, as opposed to auranofin.

Au(III) compounds are typically not very stable under biological conditions since Au(III) can be easily reduced and it has fast ligand exchange rates. Despite the limitations, many recent Au(III)-containing compounds have been rationally designed by tailoring the ligands to stabilize Au(III). The Au(III) compounds evaluated so far as ZnF inhibitors undergo reduction to Au(I) with loss of all ligands, thus it is commonly accepted that the oxidation state of incorporated gold in AuFs is 1+. By handpicking the ligands, it is possible to fine tune the stability of Au(III) complexes, stabilizing the Au(III) oxidation state even in the presence of peptides with high cysteine content such as ZnFs. Here, we explored the Au(III)(C^N) motif based on the organometallic compound $[Au(2\text{-bnpy})Cl_2]$ (2-bnpy = deprotonated 2-benzylpyridine) in comparison to a series of Au(III) complexes with typical $\kappa^2 N,N'$ chelators (2,2'-bipyridine, 4,4'-dimethyl-2,2'-bipyridine and 1,10-phenanthroline). The organometallic compound $Au(2\text{-bnpy})Cl_2$ had a *sui generis* mechanism of zinc displacement, never reported before for any metal compound, that consists in the transfer of the 2-bnpy ligand to a Cys residue from the protein, leading to Au-catalyzed C-S coupling.

Determining the actual coordination sphere of Au in the AuF obtained by interacting Au(I) and Au(III) complexes with ZnF proteins is an interesting challenge, and for that purpose, we used two innovative approaches that allowed us to obtain structural information in solution. Travelling-Wave Ion Mobility (TWIM) coupled to Mass Spectrometry (MS) was used to evaluate the interaction of the Au(I) compound $[AuCl(Et_3P)]$ with two model ZnF proteins that differ on the Zn coordination sphere: NCp7 ZnF2 (Cys_3His) and the human transcription factor Sp1 ZnF3 (Cys_2His_2). Furthermore, X-Ray Absorption Spectroscopy (XAS) was used in a "dual-probe" approach to monitor oxidation state, coordination sphere changes and geometry changes of both Au and Zn for the interaction of two series of compounds. In the first series, $[AuCl(Et\text{-}_3P)]$ and auranofin were compared when interacting with NCp7 ZnF2 and Sp1 ZnF3. TWIM-MS and XAS data indicate that $[AuCl(Et_3P)]$ leads to the formation of a Cys-Au-His AuF when interacting with Sp1 ZnF3, as opposed to the $\{Et_3PAu\}$-F species (observed by XAS) that evolves to the Cys-Au-Cys AuF (observed by TWIM-MS) identified for the interaction with NCp7 ZnF2. Finally, XAS was used to compare the interaction of two $[Au(dien)L]^{n+}$ compounds (dien = diethylenetriamine; L = Cl, n = 2; L = dmap, n = 3) with the same two zinc fingers and it was demonstrated that $[Au(dien)(dmap)]^{3+}$ retains the AuN_4 coordination sphere and square planar geometry of Au(III) when interacting with NCp7 ZnF2, suggesting that the compound behaves as a non-covalent Zn ejector. MS data confirmed Zn displacement, and led to the identification of a non-covalent adduct between $[Au(dien)(dmap)]^{3+}$ and the full-length NCp7 ZnF, supporting the non-covalent displacement hypothesis.

Preface

Zinc finger proteins were discovered recently in 1982 [1]. The first representative of this class of proteins, referred to as Transcription Factor III (TFIII), was identified by the study of *Xenopus laevis* oocytes. Zn binding was confirmed by X-ray absorption spectroscopy (XAS) [2], an alternative technique to the classical single-crystal diffraction and NMR-based techniques. This research was pioneered by Prof. Aaron Klug who combined the zinc-binding information provided by XAS with the already known sequence of TFIII and came up with the idea of a finger-like structure, which established a new principle of nucleic acid recognition. Transcription factors are proteins involved in the process of converting, or transcribing, DNA into RNA. Transcription factors include a wide number of proteins, excluding RNA polymerase, that initiate and regulate the transcription of genes. One distinct feature of transcription factors is that they have DNA-binding domains that give them the ability to bind to specific sequences of DNA called enhancer or promoter sequences. Some transcription factors bind to a DNA promoter sequence near the transcription start site and help form the transcription initiation complex [3, 4]. Regulation of transcription is the most common form of gene control. The action of transcription factors allows for unique expression of each gene in different cell types and during development.

Among the plethora of zinc fingers that have been discovered and studied so far, we chose the nucleocapsid protein (NCp7) from the human immunodeficiency virus (HIV-1) as the main protein target for the work discussed in this Thesis. One of the most remarkable structural and functional characteristics NCp7 is the presence of two -Cys-X2-Cys-X4-His-X4-Cys- zinc finger (ZnF) domains, typically found on nucleic acid binding proteins. The presence of ZnF domains is highly conserved among retroviruses [5] and any mutation on the Zn-bound residues results in the loss of biological function [6, 7]. On the infectious HIV-1, NCp7 appears closely and strongly associated to RNA on the viral core [8]. The two major functions of NCp7 are RNA binding and viral encapsidation, but new evidence also suggests that NCp7 has a role in some other processes such as RNA dimerization, Gag-Gag interactions, membrane binding, reverse transcription and stabilization of the pre-integration protein complex [5]. Inhibition of NCp7 makes the virus ineffective, so the development of NCp7 inhibitors represents an interesting alternative to the reverse transcriptase and proteases inhibition typically explored for HIV infection treatment.

The human transcription factor protein Sp1 [also known as specificity factor (1)] was also investigated as a target, with the purpose of comparing the more reactive Cys_2His_2 ZnF motif with the Cys_3His motif found in the structure of NCp7. Sp1 *per se* is also an interesting *zinc finger* target, as it is overexpressed in many cancer and it is associated with poor prognosis. Sp1's function is complex, both activating and suppressing genes essential for cancer development and tumor suppressors, in addition to genes regulating cellular proliferation, differentiation, DNA damage response, apoptosis and angiogenesis [9]. The coordination sphere has also a direct effect on the zinc affinity of the protein. NCp7's Cys_3His motif was shown to bind zinc more tightly than Sp1's Cys_2His motif [10, 11].

Many complementary approaches have been used to inactivate the NCp7 zinc fingers for therapeutic use [12, 13]. Considering the zinc center as a target, an interesting suggestion to approach selectivity is that, based on the Lewis acid considerations for different zinc coordination spheres, zinc chelators may be effective for zinc coordination spheres with a labile catalytic site whereas electrophilic attack of zinc-bound thiolate may be more favored in cases of structural zinc where there are at least 2 Cys residues in the coordination sphere [14]. This hypothesis to some extent reflects the historical situation where early attempts to inactivate the zinc center used compounds such as S-acyl 2-mercaptobenzamide thioesters [15, 16]. The cysteine residues on the NCp7 are some of the most nucleophilic of all zinc-bound thiolates in proteins [17, 18]. As such, they are substrates for alkylation by electrophiles such as maleimide and iodoacetamide. [19] In cellular assays, N-ethylmaleimide (NEM) can inhibit retroviral infectivity in a concentration-dependent manner [20]. Modification of cysteine residues by alkylating agents such as iodoacetamide, 4-vinylpyridine and acrylamide has also been used in mass spectrometric peptide mapping for protein identification [21].

Although it is considered a noble metal, gold has a very rich chemistry. It is capable of forming coordination compound in both of its typical oxidation states (1+ and 3+). Gold(I) forms linear compounds, with coordination number 2, while gold(III) form square planar compounds with coordination number 4. Electron-rich metals such as gold are particularly suitable for designing metal-based zinc ejectors because, as soft Lewis acid electrophiles, they have high affinity for Cys residues. Without the proper tridimensional folding granted by Zn coordination, NCp7 is inactivated and unable to further recognize specific nucleic acid sequences. In this work, we make use of the Lewis acid electrophilic attack strategy, based on Au(I)-phosphine compounds, gold(III)(N^N) compounds and a gold(III)(C^N) organometallic compound.

Campinas, Brazil Raphael Enoque Ferraz de Paiva

References

1. Klug, A.: The discovery of zinc fingers and their development for practical applications in gene regulation and genome manipulation. Q. Rev. Biophys. **43**(1), 1–21 (2010). https://doi.org/10.1146/annurev-biochem-010909-095056
2. Diakun, G.P., Fairall, L., Klug, A.: EXAFS study of the zinc-binding sites in the protein transcription factor IIIA. Nature, **324** (6098), 698–699 (1986). https://doi.org/10.1038/324698a0
3. Coleman, J.E.: Zinc Proteins: Enzymes, Storage Proteins, Transcription Factors, and Replication Proteins. Annu. Rev. Biochem. **61**(1), 897–946 (1992). https://doi.org/10.1146/annurev.bi.61.070192.004341
4. Emerson, R.O., Thomas, J.H., Radhakrishnan, I., Case, D., Gottesfeld, J.: Adaptive Evolution in Zinc Finger Transcription Factors. PLoS Genet. **5**(1), (2009). e1000325 https://doi.org/10.1371/journal.pgen.1000325
5. Wills, J.W., Craven, R.C.: Form, function, and use of retroviral Gag proteins. AIDS. **5** (6), 639–654 (1991). https://doi.org/10.1097/00002030-199106000-00002
6. Van Wezel, R., Liu, H., Wu, Z., Stanley, J., Hong, Y.: Contribution of the Zinc Finger to Zinc and DNA Binding by a Suppressor of Posttranscriptional Gene Silencing Contribution of the Zinc Finger to Zinc and DNA Binding by a Suppressor of Posttranscriptional Gene Silencing. J. Virol. **77**(1), 696–700 (2003). https://doi.org/10.1128/JVI.77.1.696
7. Krishna, S.S., Majumdar, I., Grishin, N.V.: Structural classification of zinc fingers. Nucleic Acids Res. **31**(2), 532–550 (2003). https://doi.org/10.1093/nar/gkg161
8. Meric, C., Darlix, J.L., Spahr, P.F.: It is Rous sarcoma virus protein P12 and not P19 that binds tightly to Rous sarcoma virus RNA. J. Mol. Biol. **173**, 531–538 (1984)
9. Beishline, K., Azizkhan-Clifford, J.: Sp1 and the 'hallmarks of cancer.' FEBS J.**282**(2), 224–258 (2015). https://doi.org/10.1111/febs.13148
10. Posewitz, M.C., Wilcox, D.E.: Properties of the Sp1 Zinc Finger 3 Peptide: Coordination Chemistry, Redox Reactions, and Metal Binding Competition with Metallothionein. Chem. Res. Toxicol. **8**(8), 1020–1028 (1995). https://doi.org/10.1021/tx00050a005
11. Rich, A.M., Bombarda, E., Schenk, A.D., Lee, P.E., Cox, E.H., Spuches, A.M., Hudson, L.D., Kieffer, B., Wilcox, D.E.: Thermodynamics of Zn 2 + Binding to Cys 2 His 2 and Cys 2 HisCys Zinc Fingers and a Cys 4 Transcription Factor Site. J. Am. Chem. Soc. **134**, 10405–10418 (2012). https://doi.org/10.1021/ja211417g
12. Mori, M., Kovalenko, L., Lyonnais, S., Antaki, D., Torbett, B.E., Botta, M., Mirambeau, G., Mély, Y. Nucleocapsid protein: A desirable target for future therapies against HIV-1. Curr. Top. Microbiol. Immunol. **389**, 53–92 (2015). https://doi.org/10.1007/82_2015_433
13. Garg, D., Torbett, B. E.: Advances in targeting nucleocapsid–nucleic acid interactions in HIV-1 therapy. Virus Res. **193**, 135–143 (2014). https://doi.org/10.1016/j.virusres.2014.07.004
14. Lee, Y., Lim, C.: Physical Basis of Structural and Catalytic Zn-Binding Sites in Proteins. J. Mol. Biol. **379**(3), 545–553 (2008). https://doi.org/10.1016/j.jmb.2008.04.004
15. Schito, M.L., Goel, A., Song, Y., Inman, J.K., Fattah, R.J., Rice, W.G., Turpin, J.A., Sher, A., Appella, E.: In Vivo Antiviral Activity of Novel Human Immunodeficiency Virus Type 1 Nucleocapsid p7 Zinc Finger Inhibitors in a Transgenic Murine Model. AIDS Res. Hum. Retroviruses. **19**(2), 91–101 (2003). https://doi.org/10.1089/088922203762688595
16. Schito, M., Soloff, A., Slovitz, D., Trichel, A., Inman, J., Appella, E., Turpin, J., Barratt-Boyes, S.: Preclinical Evaluation of a Zinc Finger Inhibitor Targeting Lentivirus Nucleocapsid Protein in SIV-Infected Monkeys. Curr. HIV Res. **4**(3), 379–386 (2006). https://doi.org/10.2174/157016206777709492
17. Maynard, A.T., Huang, M., Rice, W.G., Covell, D.G.: Reactivity of the HIV-1 nucleocapsid protein p7 zinc finger domains from the perspective of density-functional theory. Proc. Natl. Acad. Sci. U. S. A. **95**(20), 11578–11583 (1998). https://doi.org/10.1073/pnas.95.20.11578

18. Maynard, A.T., Covell, D.G.: Reactivity of Zinc Finger Cores:? Analysis of Protein Packing and Electrostatic Screening. J. Am. Chem. Soc. **123**(6), 1047–1058 (2001). https://doi.org/10.1021/ja0011616
19. Chertova, E.N., Kane, B.P., McGrath, C., Johnson, D.G., Sowder, R.C., Arthur, L.O., Henderson, L.E.: Probing the Topography of HIV-1 Nucleocapsid Protein with the Alkylating Agent N-Ethylmaleimide. Biochemistry **37**(51), 17890–17897 (1998). https://doi.org/10.1021/bi980907y
20. Morcock, D.R., Thomas, J.A., Gagliardi, T.D., Gorelick, R. J., Roser, J.D., Chertova, E.N., Bess, J.W., Ott, D.E., Sattentau, Q.J., Frank, I., et al.: Elimination of retroviral infectivity by N-ethylmaleimide with preservation of functional envelope glycoproteins. J. Virol., **79**(3), 1533–1542 (2005). https://doi.org/10.1128/JVI.79.3.1533-1542.2005
21. Sechi, S., Chait, B.T.: Modification of Cysteine Residues by Alkylation. A Tool in Peptide Mapping and Protein Identification. *Anal. Chem.* **70** (24), 5150–5158 (1998). https://doi.org/10.1021/ac9806005

Parts of this thesis have been published in the following journal articles:

de Paiva, Raphael E. F.; Du, Zhifeng; Nakahata, Douglas H.; Lima, Frederico A.; Corbi, Pedro P.; Farrell, Nicholas P. Au-catalyzed C-S aryl group transfer in Zinc Finger Proteins. *Angewandte Chemie* (International ed.), v.57, p.9305–9309, 2018.

Abbehausen, Camilla; **de Paiva, Raphael E. F.**; Bjornsnon, Ragnar; Gomes, Saulo Q.; Du, Zhifeng; Corbi, Pedro P.; Lima, Frederico A.; Farrell, Nicholas P. X-ray Absorption Spectroscopy Combined with Time-Dependent Density Functional Theory Elucidates Differential Substitution Pathways of Au (I) and Au(III) with Zinc Fingers. *Inorganic Chemistry*, v.57, p. 218–230, 2018.

de Paiva, Raphael E. F.; Du, Zhifeng; Peterson, Erica; Corbi, Pedro P.; Farrell, Nicholas P. Probing the HIV-1 NCp7 nucleocapsid protein with site-specific gold(I)-phosphine complexes. *Inorganic Chemistry,* v.56, p. 12308–1231, 2017.

de Paiva, Raphael E. F.; Nakahata, Douglas H.; Corbi, Pedro P. Synthesis and crystal structure of di-chlorido-(1,10-phenanthroline-$\kappa^2 N,N'$)gold(III) hexa-fluorido-phosphate. *Acta Cyst. E*, v.73, p. 1048–1051, 2017.

Du, Zhifeng; **de Paiva, Raphael E. F.**; Nelson, Kristina; Farrell, Nicholas P. Diversity in Gold Finger Structure Elucidated by Traveling-Wave Ion Mobility Mass Spectrometry. *Angewandte Chemie* (International ed.), v.56, p. 4464–4467, 2017.

Acknowledgements

The Ph.D. program was an interesting adventure. I thank my advisor, Prof. Pedro P. Corbi, for being extremely supportive and open to new ideas. Pedro has always been a problem-solver and that is something I admire about him. More than anything, he has been a friend for more than 7 years now, and I'm looking forward to our collaborations in the years to come! I also thank Prof. Nicholas P. Farrell for receiving me in his lab at VCU and for the opportunity to work on some very interesting projects in a very engaging working environment. Being in his lab, as part of the Science without Borders program, was one of the best experiences I had in my scientific life.

I thank Fred Lima, Saulo Gomes and Camilla Abbehausen for the collaboration on the X-Ray Absorption Spectroscopy technique. Working with you guys at the Synchrotron Facility was an extraordinary experience and I learned a lot from it. I also thank our collaborator Ragnar Björnsson for the TD-DFT calculations, which provided great theoretical support to our XAS experiments.

From LQBM at UNICAMP, I thank the lab mates Julia Nunes, Marcos Carvalho, Fernando Bergamini, Carlos Manzano and Ana Fiori for the friendly working environment. I thank Douglas Nakahata for the scientific discussions and also for the help crystallizing the Au(III) compound and refining the data collected. I'm also grateful for the opportunity to work on his Cu(II)-sulfonamide project, where we did some very interesting science combining well-known biophysical assays in a new and creative way. From LNanoMol at UNICAMP, I thank Bruno Pires for the help planning the electrochemical experiments. Finally, I thank Douglas Nakahata for carefully proofreading the draft and final version of this thesis, and Julia Nunes and Carlo Manzano for beta reading the final version.

From Farrell Lab at VCU, I thank some amazing people that I had the opportunity to work with: Zhifeng Du, one of the most competent and pro-active persons I've ever met, for all the help with MS experiments; and Erica Petterson for all the training, teachings and help with molecular biology. I thank Eric Ginsburg for our many adventures in the Richmond area. I also thank Victor Bernardes, Samantha Katner, Wyatt Johnson and James Beaton for being very receptive lab mates.

Finally, I thank our amazing lab technicians at IQ-UNICAMP: Cláudia Martelli (UV-Vis), Priscila Andrade (MS), Déborah de Alencar Simoni (X-Ray crystallography), Anderson Pedrosa and Gustavo Shimamoto (NMR). And from VCU, I thank Kevin S. Knitter (MS).

Contents

Abbreviations, Acronyms and Frequently Used Terms

"a_n" ion	Peptide fragment observed by MS/MS and CID. Corresponds to a decarbonylated "b_n" ion
apopeptide	Non-metallated protein
ATD	Arrival Time Distribution
"b_n" ion	Peptide fragment observed by MS/MS and CID. Originate from the nth amino acid from the N-terminus
bipy	2,2'-bipyridine
bnpy	2-benzylpyridine
CCDC	Cambridge Crystallographic Data Centre
CCS	Collision Cross-Section
CEM	T lymphoblast tumorigenic cell line, the associated disease is the acute lymphoblastic leukemia
CID	Collision Induced Dissociation
COSY	NMR experiment. Correlation Spectroscopy
CV	Cyclic Voltammetry
Cys	*L*-Cysteine
Cy_3P	Tricyclohexylphosphine
dien	Diethylenetriamine
DFT	Density Functional Theory calculations
dmap	4-dimethylaminopyridine
dmbipy	4,4'-dimethyl-2,2'-bipyridine
dmf	N,N-dimethylformamide
dmso	Dimethyl sulfoxide
Et_3P	Triethylphosphine
EXAFS	Extended X-Ray Absorption Fine Structure
F	Apo zinc finger protein. Appears along wiht the species that replaced Zn. One example of use is "AuF", indicating that a Au ion is coordinated to some of residues that were originally part of Zn coordination sphere in the protein
FP	Fluorescence Polarization

FWHM	Full width at half maximum
GAG	Core structural protein of HIV-1. After proteolytic processing, major viral proteins are obtained, such as the matrix (MA), capsid (CA) and nucleocapsid (NC)
HEPES	4-(2-hydroxyethyl)-1-piperazineethanesulfonic acid, an organic zwitterionic compound used as buffering agent
HMBC	NMR experiment. Heteronuclear Multiple Bond Coherence
HSQC	NMR experiment. Heteronuclear Single-Quantum Coherence
HUVEC	Human Umbilical Vein Endothelial Cells
IMS	Ion Mobility Spectrometry
MS	Mass Spectrometry
N-Ac-Cys	*N*-Acetyl-*L*-Cysteine
NC	Nucleocapsid
NCp_7	Nucleocapsid protein 7 from HIV-1
Full-length	Full-length nucleocapsid protein from HIV-1, comprising both the ZnF1 and ZnF2 domains
ZnF_2	C-terminal ZnF domain of the nucleocapid protein from HIV-1
Zn_1F	Full-length nucleocapsid protein from HIV-1 containing Zn(II) in only one of its zinc finger domains
Zn_2F	Full-length nucleocapsid protein from HIV-1 containing Zn(II) in both zinc finger domains
NEM	N-ethylmaleimide
NMR	Nuclear Magnetic Resonance
PDB	Protein Data Bank
phen	1,10-phenantholine
ROS	Reactive Oxygen Species
Sp1	Human transcription factor, comprises three ZnF domains
ZnF3	Third (C-terminal) ZnF domain of the human transcription factor protein Sp1
TD-DFT	Time-Dependent DFT
TWIM-MS	Traveling-wave Ion Mobility coupled with Mass Spectrometry
XANES	X-ray Absorption Near Edge Structure
XAS	X-ray Absorption Spectroscopy
"y_n" ion	Peptide fragment observed by MS/MS and CID. Originate from the nth amino acid from the C-terminus
ZnF	Zinc finger protein

List of Figures

List of Schemes

List of Tables

Part I
Au(I) Complexes

Chapter 1
Au(I)-Phosphine Series Design

1.1 Introduction

The zinc finger template is in general a ready source for metal ion replacement [1]. Ions such as Co(II) and Cd(II) can be used as spectroscopic probes and it is axiomatic that metal-ion replacement may also result in inhibition of function due to alterations in tertiary structure when ions preferring square-planar geometry, such as Pt(II) or Au(III), or linear geometry, such as Au(I), replace tetrahedral Zn(II). Electron-rich metals such as Au(I) are particularly suitable for designing metal-based zinc ejectors because as thiophilic Lewis acid electrophiles they have high affinity for Cys residues. Without the proper tridimensional folding, NCp7 is inactivated and unable to further recognize specific nucleic acid sequences. This strategy is also interesting because it represents an alternative to the conventional reverse transcriptase HIV-1 inhibition widely available in the clinic. Few works have reported the zinc ejecting capabilities of Au(I) compounds. Larabee et al. [2] demonstrated the potential of aurothiomalate as a zinc ejector for the specificity factor Sp1 back in 2005. Despite the particularly suitable characteristics of Au(I) for designing zinc ejectors, our group was the first to report a series of Au(I) compounds rationally designed as potent zinc ejector [3].

Compounds of general structure **[L-Au-A]^{n}** allow the modification of both the leaving ligand L and also the ancillary ligand A for fine-tuning the properties of the overall complex towards the desired biomolecular target. Phosphines and carbenes are particularly well suited to work as carrier (or ancillary) ligands when designing Au(I) complexes for they are able to form stable and strong bonds with Au(I) as well as good air and moisture stability of the formed Au(I) complexes [4, 5]. The two aromatic residues in the structure of NCp7 (Phe16 and Trp37) are responsible for π-stacking with purine and pyrimidine residues on RNA and DNA [6–8]. The importance of these residues in dictating reactivity can be explored by

---

Part of this chapter has been reproduced with permission from ACS. https://pubs.acs.org/doi/abs/10.1021/acs.inorgchem.7b01762.

R. E. Ferraz de Paiva, *Gold(I,III) Complexes Designed for Selective Targeting and Inhibition of Zinc Finger Proteins*, Springer Theses,
https://doi.org/10.1007/978-3-030-00853-6_1

introducing aromatic co-ligands with good π-stacking ability such as a purine or 4-dimethylaminopyridine (dmap) in the L position of [Au(L)(A)], leading to more efficient metal-based zinc ejectors [1]. In this chapter we explore the basicity and steric hindrance of the phosphine as well as lability and donating properties of the co-ligand (N-heterocycle vs. Cl^-) to design a series of Au(I)-phosphine complexes. This series of Au(I) compounds explores the analogy between the organic and the Lewis acid electrophiles, and we demonstrate that chemoselective auration is useful for probing the nucleocapsid topography. We also examined for comparison the "standard" gold-phosphine compound auranofin which contains a thiosugar co-ligand. The Au(I)-phosphine compounds evaluated here are shown in Fig. 1.1, along with the structures of the zinc finger proteins selected as targets.

Previous works have explored other classes of proteins as biomolecular targets for Au(I)-phosphine complexes. Glutathione-S-transferase was inhibited by auranofin [9]. [AuCl(Et_3P)] was used for targeting the de novo designed coiled coil TRIL23C, binding to Cys residues [10]. A series of Au(I)-triethylphosphine was shown to inhibit cytosolic and mitochondrial thioredoxine reductase [11]. Bis(cyanoethyl)phenylphosphine AuCl was found to selectively inhibit lymphoid tyrosine phosphatase over other tyrosine phosphatases [12]. Regarding specifically zinc finger proteins as targets, an Au(I)-triphenylphosphine with pyridylimidazole ligand was described to inhibit PARP-1 expression in A2780 cell extracts. Our group has also studied a series of Ph_3PAuL (L=Cl or N-heterocycle) compounds, and we demonstrated that metalation of NCp7 ZnF2 happens with maintenance of the phosphine ligand [3].

1.2 Experimental

1.2.1 Materials

H[$AuCl_4$], triethylphosphine (Et_3P) were purchased from Acros. Tricyclohexylphosphine, 4-dimethylaminopyridine (dmap), 2,2′-thiodiethanol and auranofin were obtained from Sigma-Aldrich. Deuterated solvents are from Cambridge isotopes or Sigma-Aldrich. The target zinc finger proteins were obtained in their non-metallated forms (apo) from Invitrogen. Protein sequences:

```
                            10         20         30         40         50
NCp7 ZnF2           KGCWKCGKEG HQMKNCTER
full-length NCp7    MQRGNFRNQR KNVKCFNCGK EGHTARNCRA PRKKGCWKCG KEGHQMKDCT ERQAN
Sp1 ZnF3            KKFACPECPK RFMSDHLSKH IKTHQNKK
```

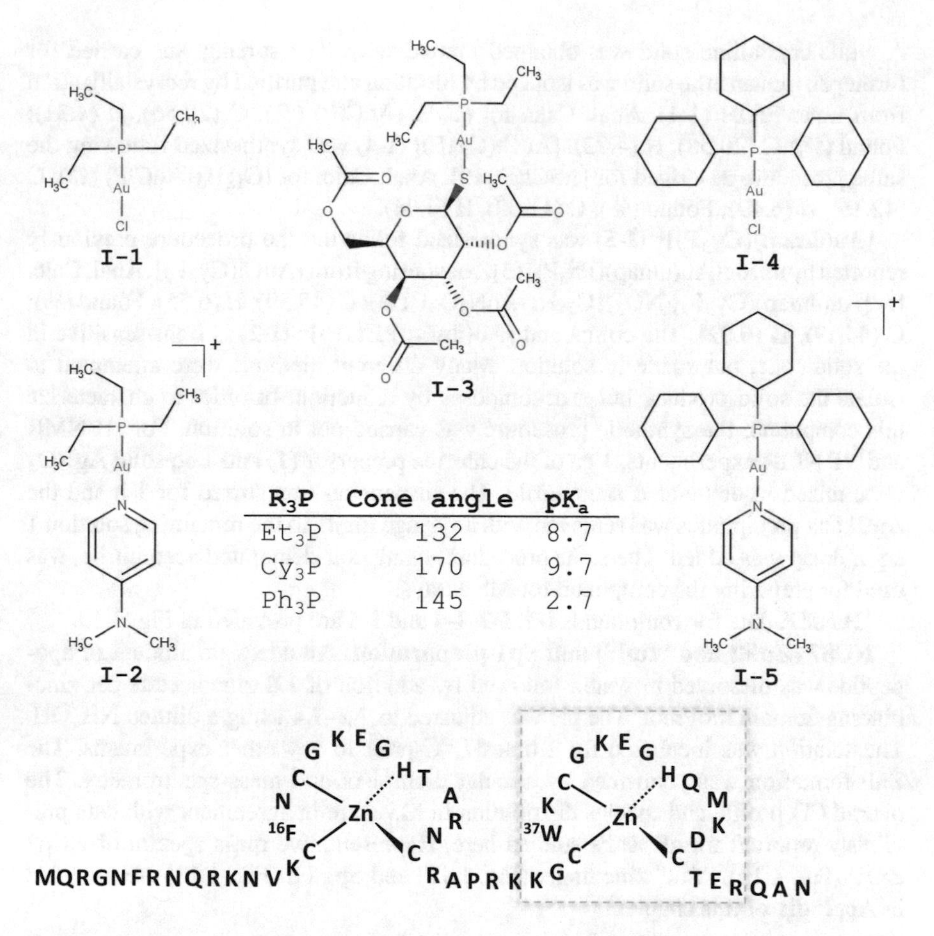

Fig. 1.1 Designed gold(I) compounds (**I-1**, **I-2**, **I-4** and **I-5**). Auranofin (**I-3**) was studied within the Et_3P series due to structural similarities. Phosphine Tolman cone angle (in degrees) and pK_a values are also given. Triphenylphosphine data (Ph_3P) is also provided as comparison. Also shown are the structures of the full-length NCp7 zinc finger (with ZnF2 boxed), highlighting the Zn-bound residues

1.2.2 Synthesis

Chloro(phosphine)gold(I). $[AuCl(Et_3P)]$ (**I-1**) was synthesized by an adaptation of a method already published [13]. Summarizing, $H[AuCl_4]$ (0.50 mmol, 194 mg) was dissolved in 1.0 mL of distilled water and the resulting solution was cooled in an ice bath. 2,2′-thiodiethanol (1.17 mmol, 117 μL) was dissolved in 360 μL of EtOH and added slowly (over ~4 h, 10 μL per addition) to the Au(III) solution. Throughout the addition, the solution changes from yellow (Au(III)) to colorless (Au(I)) and the brown-orange byproduct was filtered off. Et_3P was dissolved in 360 μL of EtOH and the resulting solution was cooled in freezer and then added to the Au(I) solution.

A white crystalline solid was obtained immediately. The stirring was carried for further 30 min and the solid was isolated by filtration and purified by recrystallization from water/EtOH (1:1). Anal. Calc. for ($C_6H_{15}AuClP$) (%): C (20.56), H (4.31); Found (%): C (20.58), H (4.22). [AuCl(Cy_3P)] (**I-4**) was synthesized following the same procedure described for [AuCl(Et_3P)]. Anal. Calc. for ($C_{18}H_{33}AuClP$) (%): C (42.16), H (6.49); Found (%): C (41.62), H (6.04).

[Au(dmap)(Cy$_3$P)]$^+$ (**I-5**) was synthesized following the procedure previously reported by us for [Au(dmap)(Ph_3P)] [3], but starting from [AuCl(Cy_3P)]. Anal. Calc. for [Au(dmap)(Cy_3P)](NO_3), $C_{25}H_{43}AuN_3O_3P$ (%): C (45.39), H (6.55). Found (%): C (44.19), H (6.02). The compound **[Au(dmap)(Et$_3$P)]$^+$** (**I-2**) is light-sensitive in the solid state, but stable in solution. Many different methods were attempted to isolate the solid product, but it decomposes by reduction. In order to characterize this compound, the synthetic procedure was carried out in solution. For ^{1}H NMR and ^{31}P NMR experiments, 1 eq of the chloride precursor (**1**) and 1 eq solid $AgNO_3$ were mixed in deuterated acetonitrile. The suspension was stirred for 1 h and the AgCl that precipitates was removed with a syringe filter. To the remaining solution 1 eq of dmap was added. The same procedure, using non-deuterated acetonitrile, was used for preparing the compound for MS assays.

^{1}H NMR data for compounds **I-1**, **I-2**, **I-4** and **I-5** are provided in Fig. 1.10.

NCp7 (ZnF2 and "full") and Sp1 preparation. An adequate amount of apo-peptide was dissolved in water, followed by addition of 1.2 zinc acetate per zinc-binding domain mol/mol. The pH was adjusted to 7.2–7.4 using a diluted NH_4OH. The solution was incubated for 2 h at 37 °C prior to any other experiments. The ZnF formation was confirmed by circular dichroism and mass spectrometry. The overall CD profile and species distribution in MS were in agreement with data previously reported for all ZnFs studied here. Representative mass spectra of NCp7 ZnF2 (Fig. 1.11), "full" zinc finger (Fig. 1.12) and Sp1 F3 (Fig. 1.13) are provided in Appendix of this chapter.

1.2.3 Ligand Scrambling Evaluation

1.2.3.1 ^{31}P NMR

A 10 mmol L^{-1} solution of compounds **I-1** to **I-5** were prepared in deuterated acetonitrile. ^{31}P NMR spectra were recorded with 256 scans at 20, 27, 34, 41 and 48 °C.

1.2.3.2 UV-Vis

The [AuCl(Cy_3P)] (4.3×10^{-4} mol L^{-1}) and [Au(dmap)(Cy_3P)]$^+$ (1.7×10^{-4} mol L^{-1}) solutions were prepared in acetonitrile. UV-Vis spectra were recorded in an HP 8453 spectrophotometer equipped with a diode array detector. The temperature was set to 20, 27, 34, 41, 48 and 55 °C using a Peltier unit. After reaching 55 °C and acquiring the respective spectra for compounds **4** and **5**, the temperature was returned to 20 °C and the spectra were recorded once again.

1.2.4 Interaction with Model Biomolecules

1.2.4.1 Ligand Displacement Induced by *N*-Acetyl-*L*-Cysteine (*N*-Ac-Cys)

NMR. 400 μL of a 10 mmol L^{-1} solution of each Au(I)-phosphine compound in CD_3CN were mixed with 400 μL of a 10 mmol L^{-1} solution of *N*-Ac-Cys in CD_3CN. ^1H and ^{31}P NMR spectra were acquired immediately after mixing and over time (after 1, 2, 24 and 96 h) for each compound in the series.

MS. A sufficiently small volume of a 14 mmol L^{-1} solution of each Au(I)-phosphine compound in acetonitrile was mixed with the same volume of a 14 mmol L^{-1} solution of *N*-Ac-Cys in water. Mass spectra were acquired immediately after mixing for each compound in the series.

1.2.4.2 Tryptophan Fluorescence Quenching Assay

7.5 mmol L^{-1} stock solutions of the dmap-functionalized compounds (**I-2** and **I-5**) were prepared. The quenchers were titrated into a cuvette containing 3.0 mL of 5 μmol L^{-1} solution of N-acetyl-tryptophan, from 1:1 up to 1:100 (*N*-Ac-Trp: quencher). Spectral window monitored: 300–450 nm, $\lambda_{ex} = 280$ nm, detector voltage = 750 V, T = 20 °C, scan rate = 600 nm/min. For the Stern-Volmer model, the linearized data was obtained at λ_{max} 362.9 nm.

1.2.5 Targeting NCp7 ZnF2 and the Full-Length NCp7 ZnF

1.2.5.1 Circular Dichroism

50 μM samples of NCp7 ZnF2, apo-NCp7 F2, full-length NCp7 ZnF or apofull-length NCp7 were used. For time-based measurements, 1.3:1 mol/mol of Au(I)-phosphine compound per ZnF core were mixed and spectra were recorded immediately after mixing and over time (after 1, 6 and 24 h). For the concentration-dependent measurements, $[Au(dmap)(Et_3P)]^+$ titrated into the ZnF solution in the molar ratio range $r_i = 0.3$–1.3.

1.2.5.2 Mass Spectrometry

1 mmol L^{-1} interaction products were obtained by mixing at room temperature a stock solution of the Au(I) complex with a ZnF stock solution (1:1 mol/mol Au complex per ZnF core in a water/acetonitrile mixture; the pH was adjusted to 7.0 with NH_4OH if needed). The final solution was incubated for up to 24 h at room temperature. MS experiments were carried out on an Orbitrap Velos from Thermo Electron Corporation operated in positive mode. Samples (25 μL) were diluted with methanol (225 μL) and directly infused at a flow rate of 0.7 μL/min using a source voltage of 2.30 kV. The source temperature was maintained at 230 °C throughout.

1.2.5.3 ^{31}P NMR

NCp7 ZnF2. 500 μL of a 0.3 mmol L^{-1} stock solution of NCp7 ZnF2 were mixed with compounds **I-1** to **I-5** dissolved in CD_3CN (1.3:1 mol/mol of gold complex per zinc finger core). Reactions were monitored from t = 0 up to 4 days. Spectra were recorded with standard phosphorus parameters and 256 scans.

Full-length NCp7 ZnF. 375 μL of a 0.129 mmol L^{-1} stock solution of full-length NCp7 zinc finger were mixed with compounds **I-1** to **I-5** dissolved in acetonitrile (1.3:1 mol/mol gold complex per zinc finger core). Reactions were monitored from t = 0 up to 8 days. The final spectra, acquired after 8 days incubation, were recorded with standard phosphorus parameters and 256 scans.

1.2.6 NCp7 (Full)/SL2 (DNA) Interaction and Inhibition

NCp7 and SL2 binding control experiment. A range of concentrations of NCp7 (aa 1–55) were mixed with 100 nM 3′-fluorescein-labeled hairpin SL2 DNA (sequence GGGGCGACTGGTGAGTACGCCCC). Samples were prepared in a 96-well black, low-binding microplate (Greiner), with total a volume of 50 μL each, containing 1.25 mmol L^{-1} NaCl, 0.125 mmol L^{-1} HEPES at pH 7.2. Fluorescence polarization (FP) readings were recorded immediately on Beckman Coulter DTX880 plate reader.

Inhibition of NCp7-SL2 binding experiment. The compounds at various concentrations were incubated with 5 μmol L^{-1} full-length NCp7 ZnF for 1 h followed by addition of SL2 DNA (final concentration of 100 nmol L^{-1}). Samples were prepared in a 96-well black, low-binding microplate (Greiner), with a total volume of 50 μL each, containing 1.25 mmol L^{-1} NaCl, 0.125 mmol L^{-1} HEPES at pH 7.2. Fluorescence polarization (FP) readings were recorded immediately on Beckman Coulter DTX880 plate reader. The full-length NCp7 ZnF concentration was chosen from the protein-SL2 binding experiment such that 90% of SL2 was bound. FP readings were recorded immediately after addition of the SL2 DNA on Beckman Coulter DTX880 plate reader.

1.3 Results and Discussion

1.3.1 Ligand Scrambling Assessed Using Variable Temperature Experiments

Ligand scrambling has been extensively reported for Au(I)-phosphine complexes [3, 14, 15], so we dedicated some attention to evaluate what is the influence of this phenomenon in our system. The following reaction can take place in solution, and justifies some of the major species observed in the mass spectra of the Au(I)-phosphine compounds:

$$2[(\boldsymbol{R_3P})\boldsymbol{AuL}]^{+} \rightarrow \left[(\boldsymbol{R_3P})_2\boldsymbol{Au}\right]^{+} + [\boldsymbol{AuL_2}]^{+}$$

The extent of ligand scrambling depends on intrinsic (steric hindrance and basicity of phosphine, as well as nature of the co-ligand) and extrinsic (polarity of the solvent, ionic strength of the medium) factors [14]. Here we evaluated only the intrinsic factors, with the extrinsic factors kept constant (acetonitrile as solvent, without added electrolytes). In order to evaluate to which extent this reaction is important in our systems, some temperature-dependent experiments were done: ^{31}P NMR and UV-Vis spectroscopy. In both experiments, increasing the temperature shifts the equilibrium towards the products while decreasing the temperature can make the reaction return to its initial state.

Spectral changes were observed when running temperature-dependent UV-Vis experiments (Fig. 1.14). The major changes observed for [AuCl(Cy_3P)] happen with the bands centered at 236 and 241 nm. Coalescence and intensity increase were observed with temperature increase. For [Au(dmap)(Cy_3P)]$^+$, the presence of a pyridine derivative ligand was also useful for assessing the system using UV-Vis spectroscopy. The band centered at 216 nm increased in intensity with temperature increase, while the ligand-related transition at 282 nm had an intensity decrease. For both compounds, returning the temperature to 20 °C led to identical spectra as the ones acquired initially at 20 °C. Furthermore, isosbestic points were identified with the temperature increase for [AuCl(Cy_3P)] and [Au(dmap)(Cy_3P)]$^+$ (at 242 and 260 nm respectively). That is indicative that an equilibrium is taking place in solution, consistent with the ligand scrambling reaction proposed.

The ^{31}P chemical shift of compound [AuCl(Et_3P)] suffers a larger temperature effect (Fig. 1.15, $\Delta\delta$ of 0.37 ppm) when compared to [AuCl(Cy_3P)] (Fig. 1.16, $\Delta\delta < 0.01$ ppm). The co-ligand L also plays a role on the ligand scrambling reaction. The compound [Au(dmap)(Cy_3P)]$^+$ underwent a slightly more prominent phosphorus signal shift (Fig. 1.17, $\Delta\delta$ of 0.457 ppm) with the temperature increase than its chloride precursor. Combining the data acquired by both techniques, it is possible to affirm that ligand scrambling happens to only a small extent in our compounds.

1.3.2 Interaction with Model Biomolecules

N-Ac-Cys was used as model for the interaction of Au(I)-phosphine series with sulphur-containing proteins. The interaction was followed by MS and ^{31}P NMR.

For the chloride precursors (compounds **I-1** and **I-4**) a sulphur-bridged [R_3PAu-(*N*-Ac-Cys)-$AuPR_3$] species appears as the most abundant peak. These species appear at 792.13 and 1116.42 m/z respectively for compounds **I-1** and **I-4** (Figs. 1.18 and 1.20). Simple chloride replacement by *N*-Ac-Cys is also observed for compound **I-4** (Fig. 1.20), at 640.23 m/z corresponding to [Cy_3PAu-(*N*-Ac-Cys)]. A trinuclear gold species is observed for **I-1** after 10 min incubation with *N*-Ac-Cys, at 1151.13 m/z corresponding to $[Au_3(N\text{-Ac-Cys})_2(Et_3P)_2]^+$, but this species is no longer present after 24 h incubation (Fig. 1.20b). For the dmap-functionalized compounds (**I-2** and **I-5**), as opposed to the Cl precursors, unreacted $[R_3PAu]^+$ species are observed. When compound **I-2** interacts with *N*-Ac-Cys (Fig. 1.19), the ligand scrambling bisphosphine $[Au(Et_3P)_2]^+$ product is the most abundant peak, at 433.14 m/z. Simple ligand replacement $[Et_3PAu\text{-}(N\text{-Ac-Cys})]^+$ and the S-bridged species $[Et_3PAu\text{-}(N\text{-Ac-Cys})\text{-}AuPEt_3]^+$ are also observed. For compound **I-5** (Fig. 1.21), the most abundant Au-containing peaks correspond the unreacted $[Cy_3PAu]^+$ species and also to the bisphosphine compound $[Au(Cy_3P)_2]^+$ at 757.42 m/z. The presence of the unreacted $[R_3PAu]^+$ peaks are indicative of a slower reaction rate for compounds **I-2** and **I-5** when compared to the chloride precursors **I-1** and **I-3**.

When **I-1** and **I-4** were mixed with *N*-Ac-Cys under the experimental conditions described for the ^{31}P NMR experiment, precipitation was observed immediately upon mixture (concentrations were higher than that used for the MS experiments previously described). This can be explained by the possible polymeric nature of the product, given the sulphur-bridged bimetallic species observed for those compounds by MS. Furthermore, the Au(I)(*N*-Ac-Cys) compound was synthesized before and it is insoluble in most common solvents, being slightly soluble only in dmso [16]. On the other hand, when compounds **I-2** and **I-5** were mixed with *N*-Ac-Cys, the product remained in solution allowing us to further characterize the system. For the interaction of $[Au(dmap)(Et_3P)]^+$ with *N*-Ac-Cys, a shift in the phosphorous signal was observed from 30.05 to 47.24 ppm, while for $[Au(dmap)(Cy_3P)]^+$ the shift was from 51.82 to 57.87 ppm (Fig. 1.22). The downfield shift is consistent with dmap replacement by a sulfur-donor.

In order to first explore the π-stacking capabilities of the dmap-functionalized compounds (**I-2** and **I-5**), a tryptophan fluorescence quenching assays was performed. Using the Stern-Volmer model (Fig. 1.23), the association constants (K_{SV}) obtained for $[Au(dmap)(Et_3P)]^+$ and $[Au(dmap)(Cy_3P)]^+$ were 4.0×10^5 and 6.2×10^5 respectively. The straight-line shape of the Stern-Volmer plots for both systems is indicative of pure static quenching, as expected for an interaction based on π-stacking. This data models the first step of association between compounds **I-2** and **I-5** and a tryptophan-containing protein, as NCp7.

Table 1.1 ^{31}P NMR peaks summary for the reaction products identified after the interaction of Au(I)-phosphine compounds with NCp7 ZnF2 and 'full' NCp7

Compound	δ^{31} P/ppm		
	Reagent	NCp7 ZnF2 products ($\Delta\delta$)	"Full" NCp7 product ($\Delta\delta$)
[AuCl(Et_3P)]	33.63	43.04 (9.41); 49.62 (15.99)	47.28 (13.65)
[Au(dmap)(Et_3P)]$^+$	30.05	46.96 (16.91)	47.28 (17.23)
Auranofin	39.05	43.04 (3.99); 34.67 (−4,39)	39.17 (0.12); 32.40 (−6.65)
[AuCl(Cy_3P)]	55.42	56.54 (1.12); 59.80 (4.37)	64.12 (8.70)
[Au(dmap)(Cy_3P)]$^+$	51.82	57.22 (5.40)	64.11 (12.29)

N-Ac-Cys products: [Au(dmap)(Et_3P)]$^+$ 47.24 ppm; [Au(dmap)(Cy_3P)]$^+$ 57.67 ppm

1.3.3 Targeting NCp7 (ZnF2)

1.3.3.1 ^{31}P NMR

The reaction of NCp7 (ZnF2) with the different gold(I) compounds was followed over time by ^{31}P NMR spectroscopy. Phosphorus NMR is a technique sensitive to nature of the L ligand in a R_3P-Au-L compound. When interaction with a biomolecule, His versus Cys coordination can be easily distinguished by ^{31}P NMR. When a Cl or N-heterocyclic ligand is replaced by sulfur, deshielding is often observed on the ^{31}P signal. Table 1.1 summarizes the data obtained when targeting NCp7 (ZnF2).

After reaction with NCp7, compounds **I-1**, **I-2**, **I-4** and **I-5** presented a downfield shift in the phosphorus signal (Fig. 1.2), indicative of Cys-binding. Two phosphorous-containing species were observed for the chloride precursors [AuCl(Et_3P)] and [AuCl(Cy_3P)], indicating a fast reaction rate and lack of selectivity. On the other hand, the dmap-functionalized compounds formed a single phosphorous-containing signal, which is indicative of a slower reaction rate with the target protein as consequence of replacement of the labile ligand chloride by the stronger sigma donor dmap.

Auranofin, on the other hand, has a unique behavior among the compounds studied here. A more shielded phosphorous signal appears upon interaction with NCp7, indicative of a possible His-binding.

1.3.3.2 ESI-MS

The ESI-MS spectrum of the reaction between compound [AuCl(Et_3P)]$^+$ and NCp7 (ZnF2) immediately upon incubation shows a number of Au-peptide adduct peaks (Fig. 1.3a). The presence of the {(Et_3P)Au} moiety bound to peptide, as indicated by

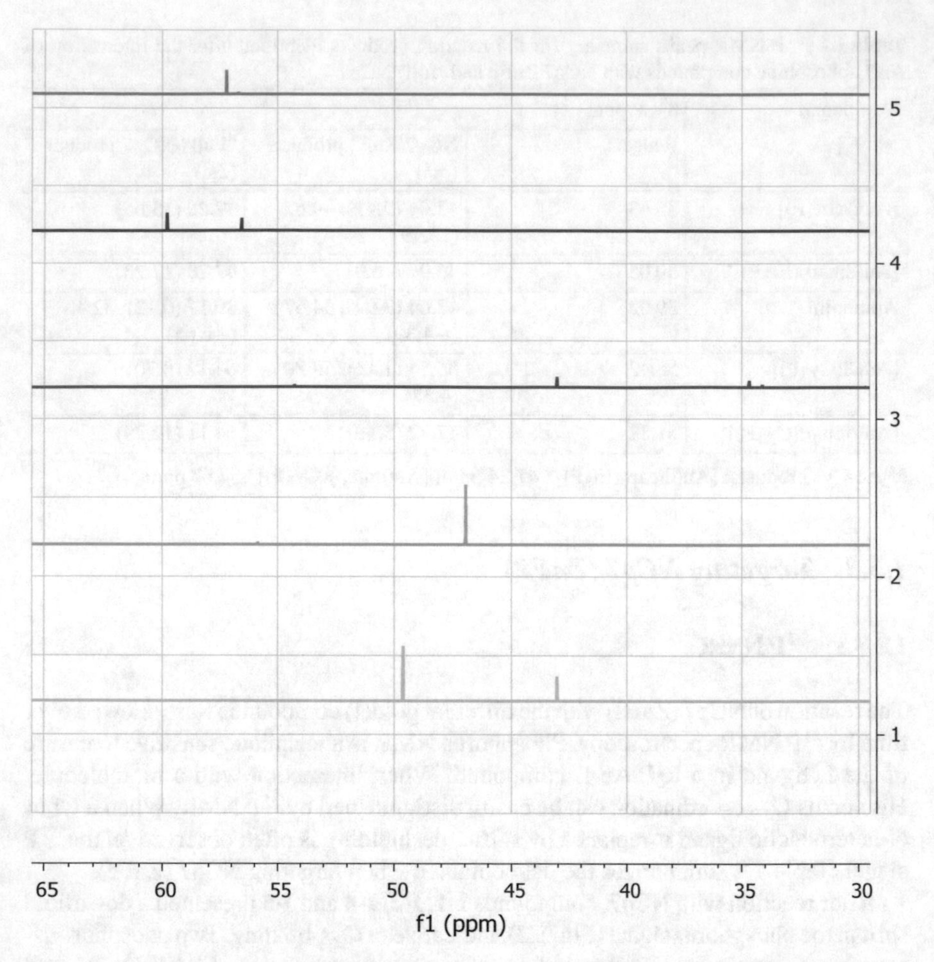

Fig. 1.2 ^{31}P NMR spectra of the reaction product obtained after incubation of NCp7 ZnF2 with Au(I)-phosphine compounds for 6 h. (**1**) [AuCl(Et$_3$P)], (**2**) [Au(dmap)(Et$_3$P)]$^+$, (**3**) auranofin, (**4**) [AuCl(Cy$_3$P)] and (**5**) [Au(dmap)(Cy$_3$P)]$^+$

^{31}PNMR data, is confirmed by the presence of major peaks attributed to Au$_2$(Et$_3$P)-apoNCp7 and the oxidized Au(Et$_3$P)-apoNCp7 species, caused by zinc displacement, along with free peptide (apoNCp7) and Au(Et$_3$P)-NCp7 ZnF2. The most prominent peaks are those at 828.62 (3+) and 1242.43 (2+) corresponding to the Au-ZnF species while the AuF (Au-apoF) species are also observed as minor species. At longer reaction times (2, 6 and 24 h), the spectra are quite similar, indicating the reaction between [AuCl(Et$_3$P)]$^+$ and NCp7 (ZnF2) happens very fast. A different situation occurs for the dmap derivative, [Au(dmap)(Et$_3$P)]$^+$ (compound **I-2**) where the auration of peptide is much slower. Immediately upon incubation with NCp7 (ZnF2), compound **I-2** remains unreacted and only ZnF signals are observed (Fig. 1.3b). Zinc displacement is only observed after 6 h, with signals assigned to apo-peptide identified. Au incorporation into the structure of the protein is only observed after 24 h. Signals assigned to $[apoNCp7\text{-}AuPEt_3]^{4+}$ at 630.51 and AuF^{3+} at 806.65 m/z were the very first Au-containing peptide species observed for this compound, clearly indicating the much slower reaction rate of $[Au(dmap)(Et_3P)]$ when targeting NCp7 (ZnF2). The presence of the thiosugar in the structure of auranofin (**I-3**) also slows down the reaction rate with NCp7 (ZnF2). For this compound, no Au adducts are observed immediately upon incubation but after 2 h the species $[NCp7\text{-}Au(PEt_3)]^{3+}$ at 867.99 m/z was identified (Fig. 1.24). Auranofin binding to HIV-2 nucleocapsid protein from two different isolates was recently studied [17]. Au(I) and Au(I)-Et_3P adducts to the target peptides were observed after 1 h of incubation. As the reaction evolved, AuF signals became more abundant for both peptides evaluated. Furthermore, auranofin interaction with the model Cys_2His_2 zinc finger PYKCPECGKSFSQKSDLVKHQRTHTG was also recently reported [18]. Auranofin was not able to displace Zn and no Au(I) adducts were observed.

When the carrier ligand Et_3P is replaced by Cy_3P, a similar overall reactivity trend is observed. The Cl-containing precursor (compound **I-4**) is much more reactive than the dmap functionalized compound (**I-5**). On the other hand, the aurated species identified are slightly different. For $[AuCl(Cy_3P)]^+$, immediately upon incubation the following Au-containing species and their respective m/z were identified: $Au\text{-}ZnF^{3+}$ at 828.62, [oxidized apoNCp7-$\{Au(Cy_3P)\}]^{3+}$ at 900.39, $[ZnF\text{-}AuCy_3P]^{3+}$ at 922.03, $[apoF\text{-}Au_2Cy_3P]^{3+}$ at 966.38 and $Au\text{-}ZnF^{2+}$ at 1242.43 (Fig. 1.3c). $[Au(dmap)(Cy_3P)]$ (**I-5**), as opposed to (**I-2**), reacts immediately with NCp7 (Fig. 1.3d), with a few Au-containing species appearing at t = 0. Some examples are $Au\text{-}ZnF^{3+}$ at 828.62, $[oxiapoF\text{-}AuCy_3P]^{3+}$ at 900.39, $[apoF\text{-}Au_2Cy_3P]^{3+}$ at 966.36 and $Au\text{-}ZnF^{2+}$ at 1242.43.

Based on the temporal evolution of Au-containing species, the following reactivity trend can be proposed for the Au(I)-phosphine series with NCp7 (ZnF2) based on ESI-MS data:

$$[\mathrm{AuCl(Et_3P)}] = \left[\mathrm{AuCl}\left(\mathrm{Cy_3P}\right)\right](\mathrm{t}=0)^3\left[\mathrm{Au(dmap)}\left(\mathrm{Cy_3P}\right)\right]^+(\mathrm{t}=0,\ \text{few species}) > \text{auranofin}(\mathrm{t}=0,\ \text{few species}) > \left[\mathrm{Au(dmap)(Et_3P)}\right]^+(24\ \text{hours})$$

The most reactive compounds were the Cl-containing precursors, indicating that the lability of the co-ligand is the first aspect that needs control to fine tune the overall reactivity of the compounds. After that, the bulkiness of the phosphine ligands plays a role, with Cy_3P-containing compounds reacting faster than the Et_3P-analogs. With respect to the electronic factor, Tolman [19] has shown that Et_3P and Cy_3P have about the same electron donating capabilities, with Et_3P being slightly more withdrawing. That can be better observed by comparing the Au–P and Au-Cl distances found in the structures of $[AuCl(Cy_3P)]$ and $[AuCl(Et_3P)]$ (CCDC entries BOPLIB10 and SATTEM respectively): Au–P 2.242 and 2.232 Å, Au-Cl 2.279 and 2.306 Å. The electronic effect alone does not explain the reactivity observed, and for that reason a combination of good σ-donating capabilities and π acidity of the co-ligand dmap, in addition to the properties of Et_3P might be the explanation for the low reactivity

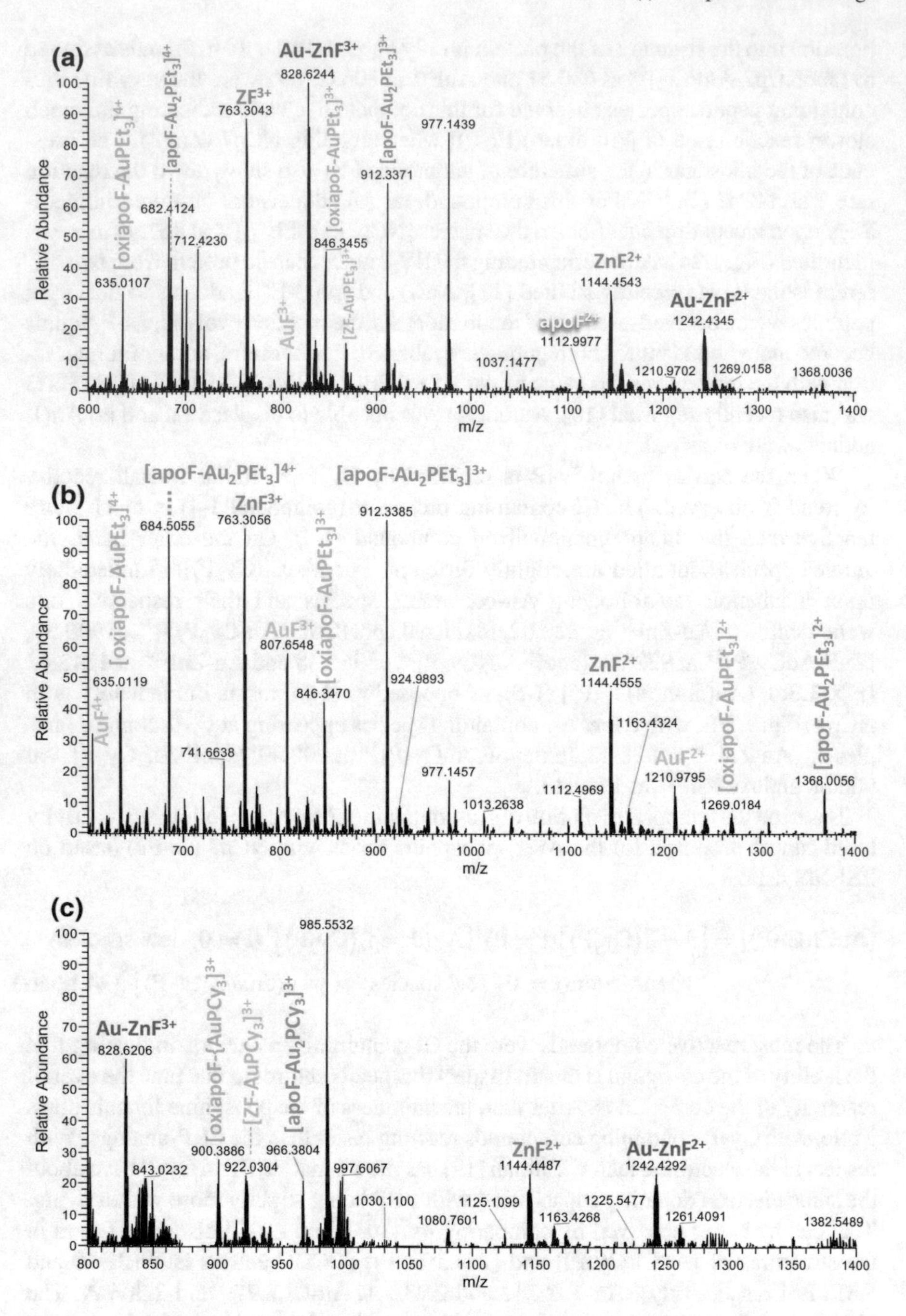

Fig. 1.3 ESI-MS spectra (positive mode) for the reaction between NCp7 (ZnF2) and **a** $[AuCl(Et_3P)]$, **b** $[Au(dmap)(Et_3P)]^+$, **c** $[AuCl(Cy_3P)]$ and **d** $[Au(dmap)(Cy_3P)]^+$. (**a**), (**c**) and (**d**) were obtained immediately after incubation. (**b**) was obtained after 6 h of incubation. Reoccurring species are color-coded across the spectra

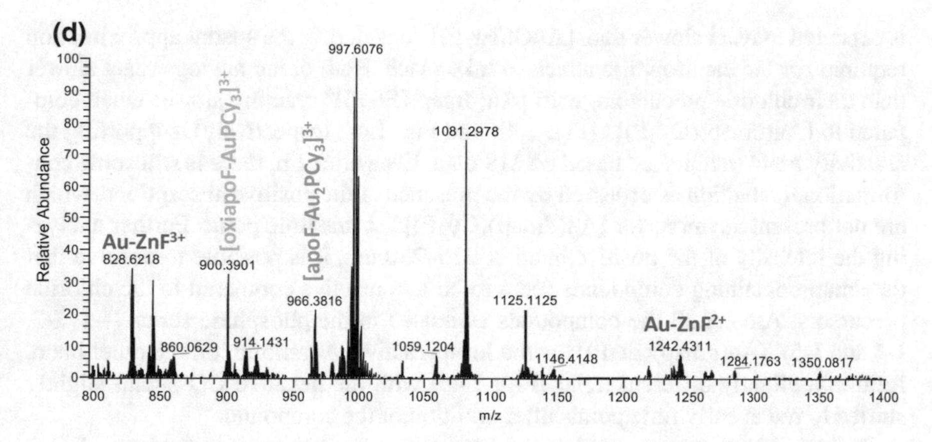

Fig. 1.3 (continued)

of $[Au(dmap)Et_3P]^+$ compared to the other compounds synthesized here. The effect of the co-ligand is further observed in the case of auranofin, unique among the Au(I) compounds evaluated here since His-binding is observed and further confirmed based on MS/MS data (Fig. 1.25).

1.3.3.3 Circular Dichroism

Circular dichroism is a useful spectroscopic technique that can be used to evaluate conformational changes caused by zinc displacement. NCp7 (ZnF2) characteristic CD profile has a negative signal centered around 205 nm, which is shifted to 200 nm upon zinc displacement. Furthermore, the most prominent change in the CD spectrum of the apo-peptide, when compared to NCp7(ZnF2), is the loss of the positive absorptions at ~190 nm and in the range 215–225 nm. As the CD profile of NCp7 (ZnF2) differs extensively from that of the apo-peptide, CD can be used to follow the time-dependent zinc displacement caused by the reaction with gold(I)-phosphine compounds.

All compounds evaluated in this study cause some degree of conformational change on NCp7 (ZnF2) secondary structure upon interaction. It is possible to observe a decrease in the intensity of the positive bands and a slight blue shift for the negative one when NCp7 is incubated with $[AuCl(Et_3P)]$, $[Au(dmap)(Et_3P)]^+$, $[AuCl(Cy_3P)]$ and $[Au(dmap)(Cy_3P)]^+$. These observations are indicative of major conformational changes that follow a time-dependent mechanism. Analyzing the intensity of the positive band at 215–230 nm, $[AuCl(Et_3P)]$ seems to react significantly faster than the other evaluated compounds, followed by $[AuCl(Cy_3P)]$ (Fig. 1.4a and Fig. 1.4c respectively). This trend can be explained considering the lability of the Cl^- ligand and also the cone angle of the phosphine ligand. Cy_3P is bulkier compared to Et_3P (Tolman cone angle of 170° vs. 132°) so $[AuCl(Cy_3P)]$

is expected to react slower than [$AuCl(Et_3P)$] considering the spatial approximation required for the electrophilic attack to take place. Both dmap analogs react slower than their chloride precursors, with [$Au(dmap)(Et_3P)$]$^+$ reacting slower when compared to [$Au(dmap)(Cy_3P)$]$^+$ (Fig. 1.4b and Fig. 1.4d, respectively), supporting the reactivity trend established based on MS data. Even after 1 h, there is still some conformational retention as observed by the presence of the positive absorptions, which are not present anymore for [$Au(dmap)(Cy_3P)$]$^+$ at this time point. Further analyzing the intensity of the positive band at 215–230 nm, it is possible to observe that the dmap-containing compounds seem to be less reactive compared to the chloride precursors. Among all the compounds evaluated in the phosphine series (**I-1**, **I-2**, **I-4** and **I-5**), [$Au(dmap)(Et_3P)$]$^+$ is the least reactive. Auranofin, on the other hand, follows a slightly different trend (Fig. 1.4c) with the negative CD signal slightly shifted to red at early time points after addition of the compound.

As a complementary measurement, [$Au(dmap)(Et_3P)$]$^+$ was titrated into a NCp7 sample (50 μM; $r_i = 0, 0.3, 0.5, 0.8, 1.0$ and 1.3) and CD spectra were acquired

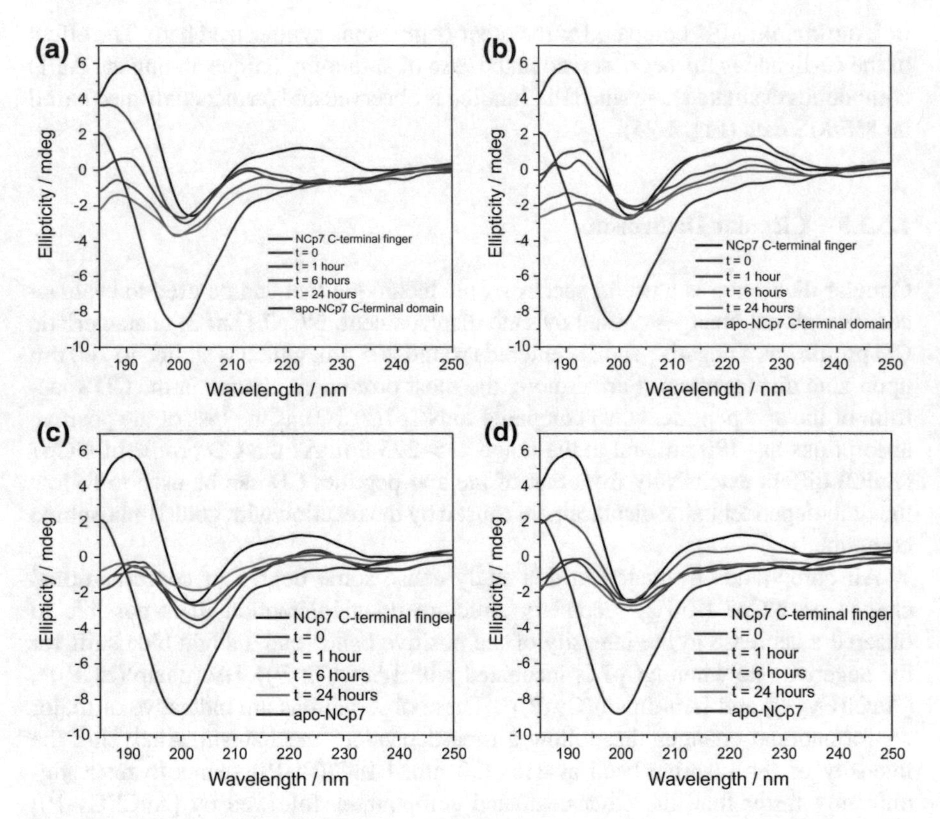

Fig. 1.4 **a–e** Time-dependent CD spectra following the reaction between the indicated Au(I)-phosphine compound and NCp7 C-terminal finger (**a–e** [$AuCl(Et_3P)$], [$Au(dmap)(Et_3P)$]$^+$, auranofin, [$AuCl(Cy_3P)$] and [$Au(dmap)(Cy_3P)$]$^+$ respectively). The titration of [$Au(dmap)(Et_3P)$]$^+$ into NCp7 (ZnF2) was also followed by CD as shown in (**f**)

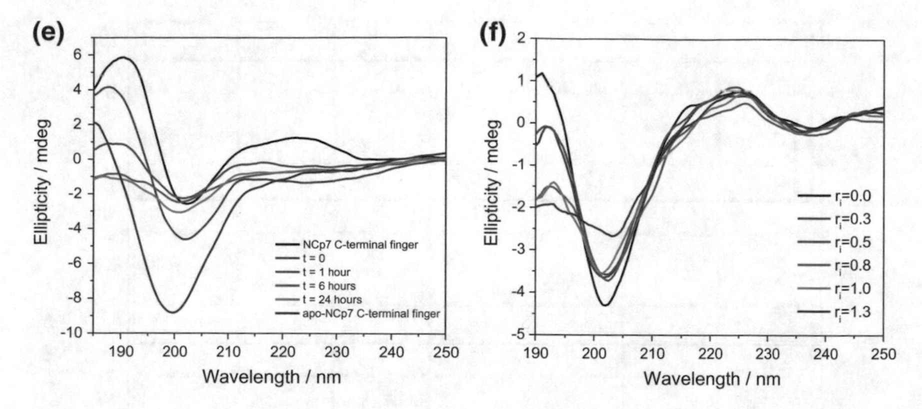

Fig. 1.4 (continued)

immediately after mixing. Figure 1.4f shows a concentration dependent conformational change induced by $[Au(dmap)(Et_3P)]^+$, with an isodichroic point at 198 nm. The positive band at 215–230 nm is maintained throughout the experiment and the presence of the isodichroic point suggests a 2-species equilibrium in solution. Combining these two pieces of information, it is possible to suggest that the ZnF structure is not completely disrupted but some conformational changes are caused by auration.

1.3.4 Targeting the Full-Length NCp7 Zinc Finger

The reactivity of the Au(I)-phosphine series was further explored by targeting the viral full-length NCp7 zinc finger. The full-length NCp7 ZnF relies on the cooperation between the two zinc finger domains, which appear linked by a highly basic sequence RAPRKKG, for recognizing specific nucleic acid sequences.

1.3.4.1 ^{31}P NMR

The interaction of the Au(I)-phosphine series with full-length NCp7 ZnF was first investigated by ^{31}P NMR, using the same approach followed for NCp7 ZnF2. Figure 1.5 shows the spectra obtained for the reaction products after 8 days incubation at 37 °C. Figures 1.26, 1.27, 1.28, 1.29 and 1.30 compare the spectra obtained for the reaction products with NCp7 ZnF2 and full-length ZnF for each compound and Table 1.1 summarizes the data discussed here.

The full-length NCp7 zinc finger has a different behavior when compared to ZnF2. A single ^{31}P signal was identified for the reaction product of this protein with the designed Au(I)-phosphine compounds. On the other hand, auranofin produced two product peaks. Auranofin also differs from the other compounds by being the only

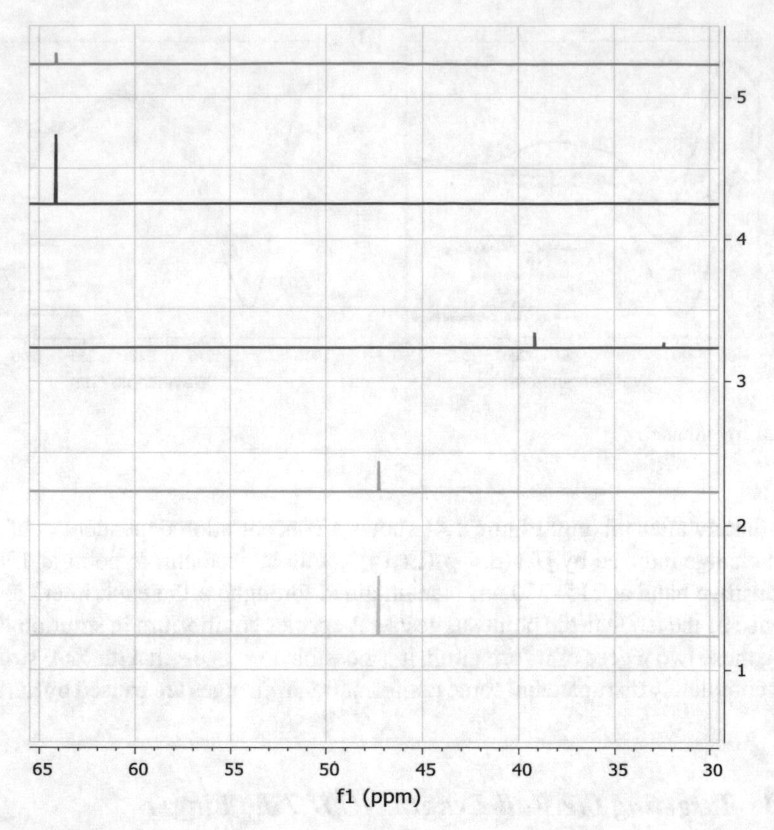

Fig. 1.5 ^{31}P NMR spectra for the reaction products obtained after incubation of full-length NCp7 zinc finger with Au(I)-phosphine compounds for 8 days. (**1**) [AuCl(Et_3P)], (**2**) [Au(dmap)(Et_3P)]$^+$, (**3**) auranofin, (**4**) [AuCl(Cy_3P)] and (**5**) [Au(dmap)(Cy_3P)]$^+$

compound that leads to the formation of a more shielded ^{31}P species upon interaction with the protein, in agreement with the data obtained for NCp7 ZnF2, and suggesting a His-binding. Furthermore, the ^{31}P signals observed here do not perfectly match the ones observed when interacting the Au(I)-phosphine compounds with ZnF2. This is indicative that the phosphine ligand appears in different chemical environments when interacting with both targets (ZnF2 and "full" zinc finger). The only case where the chemical shift of the reaction products with both targets is almost identical is for [Au(dmap)(Et_3P)]$^+$ (Fig. 1.27), another evidence that a slower reaction rate guides the reaction towards the formation of most stable Au(I)-protein bond.

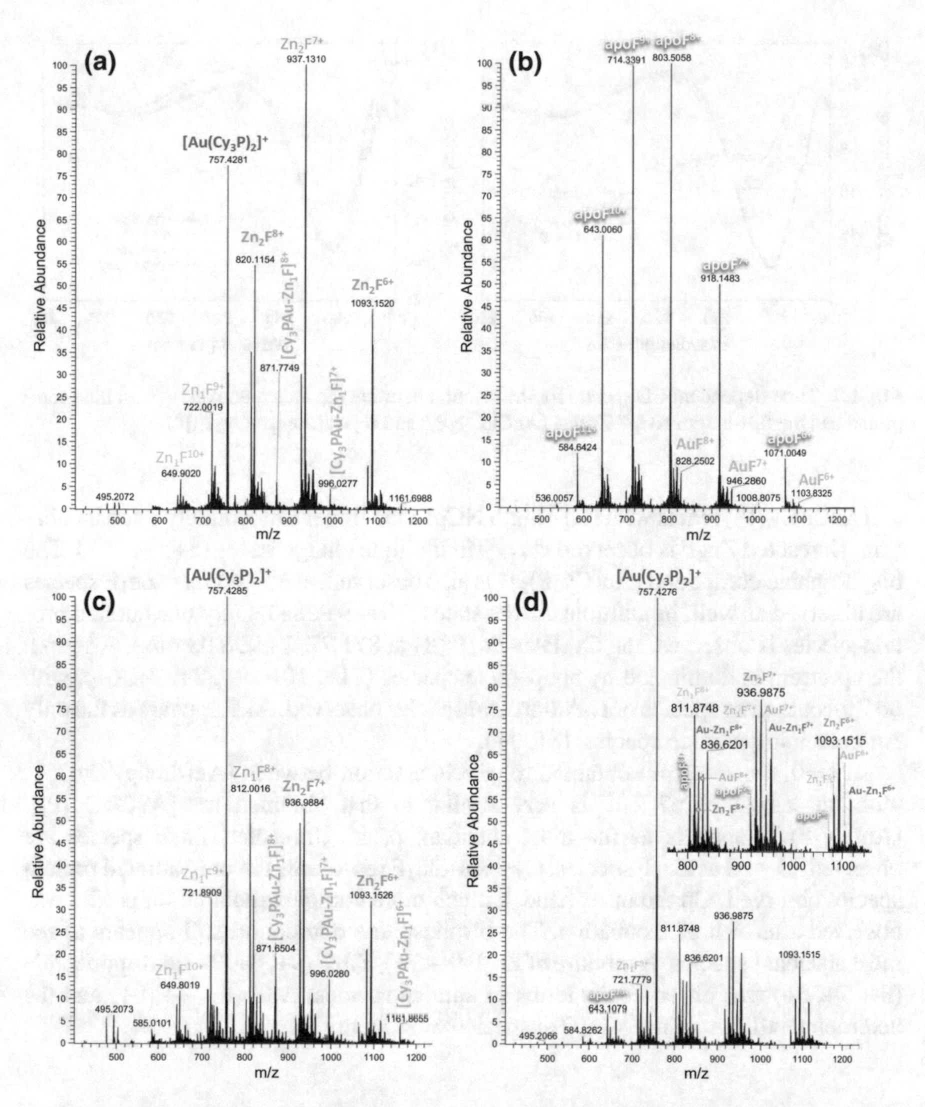

Fig. 1.6 ESI-MS spectra (positive mode) for the reaction between the full-length NCp7 ZnF and **a** $[AuCl(Cy_3P)]$ immediately after incubation, **b** $[AuCl(Cy_3P)]$ after 6 h incubation, **c** $[Au(dmap)(Cy_3P)]^+$ immediately after incubation and **d** $[Au(dmap)(Cy_3P)]^+$ after 6 h incubation. Reocurring species are color-coded

1.3.4.2 ESI-MS

The Cy_3P series, comprising compounds $[AuCl(Cy_3P)]$ and $[Au(dmap)(Cy_3P)]^+$, was selected for further investigation, now targeting the dinuclear full-length NCp7 zinc finger. The MS spectra obtained for the interactions are shown in Fig. 1.6.

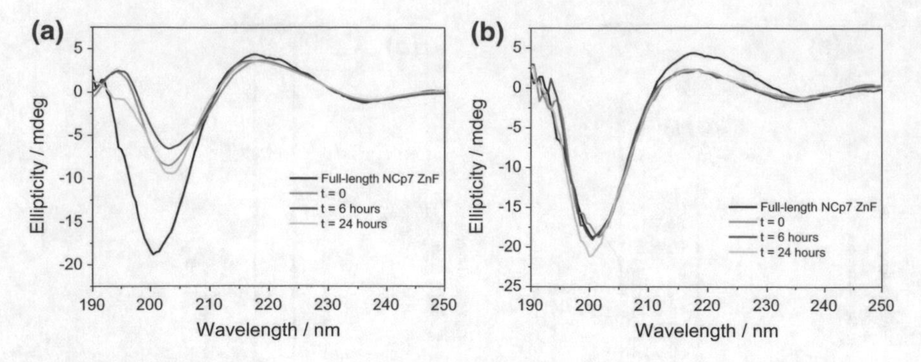

Fig. 1.7 Time-dependent CD spectra for the reaction between the indicated Au(I)-phosphine compound and the full-length NCp7 ZnF. **a** [AuCl(Cy_3P)] and **b** [Au(dmap)(Cy_3P)]$^+$)

[AuCl(Cy_3P)] reacts with full-length NCp7 zinc finger immediately upon incubation. Unreacted Zn_2F is observed at t = 0 in multiple charge states (8+, 7+, 6+). The bisphosphine compound [Au(Cy_3P)$_2$]$^+$ is also observed, at 757.43 m/z. Zn_1F species are observed as well, in multiple charge states (10+, 9+, 8+). Only one aurated protein species is observed, the Cy_3PAu-Zn_1F (8+ at 871.77, 7+ 996.03 m/z). After 6 h the spectrum is dominated by apopeptide species (11+, 10+, 9+, 8+, 7+, 6+), with no Zn-containing species or L-Au-protein species observed. AuF appears as the only Au-containing peptide species, (8+, 7+).

At t = 0, the spectrum obtained for the interaction between [Au(dmap)(Cy_3P)]$^+$ with full-length NCp7 ZnF is very similar to that obtained for [AuCl(Cy_3P)]. [Au(Cy_3P)$_2$]$^+$ appears as the most abundant peak. Unreacted Zn_2F species are observed, as well as Zn_1F species. Cy_3PAu–Zn_1F represents the only aurated protein species observed. On the other hand, a much more complex metallation profile was observed after 6 h of incubation. The bisphosphine compound still appears as the most abundant species. A mixture of Zn_2F (8+,7+), Zn_1F (9+, 8+, 7+) and apopeptide (8+, 7+, 6+) was observed. In terms of aurated species, AuF (8+, 7+, 6+) and the heterobimetallic Au-Zn_1F (8+, 7+) species were identified.

1.3.4.3 Circular Dichroism

The full-length NCp7 ZnF has a CD profile very similar to that observed for NCp7 ZnF2, with a valley at 200 nm and a positive absorption region with a maximum at 218 nm. The interaction of the Cy_3P series with full-length NCp7 ZnF was further investigated (Fig. 1.7). Interestingly, a quick reaction (at t = 0) is observed for compound **I-4**, with an increase in intensity of the negative absorption at 200 nm. On the other hand, the dmap-functionalized compound **I-5** induced slight changes on the positive band and has almost no effect on the negative band of full-length NCp7 ZnF, even after 24 h incubation.

1.3.5 Interaction with Sp1 ZnF3

The interaction of the Au(I)-phosphine series with Sp1 (ZnF3) was followed by mass spectrometry. Sp1 (ZnF3) was shown to be intrinsically more reactive than another zinc finger, the viral HIV-1 NCp7 (ZnF2). The compound $[AuCl(Et_3P)]^+$ reacts immediately with Sp1 (ZnF3), giving a clean spectrum containing AuF only (Fig. 1.8), with the species AuF^{6+} (594.96), AuF^{5+} (713.75), AuF^{4+} (891.93) and AuF^{3+} (1188.91) being the most abundant peaks in the spectrum and very few other peaks observed. This is indicative of a mechanistically well-defined reaction of Zn displacement and Au incorporation. $[Au(dmap)(Et_3P)]$, opposed to what is observed when targeting NCp7 (ZnF2), also reacts fast with Sp1 (ZnF3). Many AuF-related species are observed as well, such as AuF^{6+} (594.96), AuF^{5+} (713.75), AuF^{4+} (891.93) and AuF^{3+} (1188.91). As opposed to $[AuCl(Et_3P)]^+$, the heterobimetallic species Au-ZnF is also observed for the interaction of $[Au(dmap)(Et_3P)]$ with Sp1 (ZnF3). To some extent, peptide oxidation is also observed, given the presence of the oxidized species $oxiF^{5+}$ (674.15) and $oxiF^{4+}$ (842.44).

Auranofin (**3**) interacts with Sp1 (F3) producing a much wider variety of Au-containing species immediately after incubation (Fig. 1.24). AuF species were identified, such as AuF^{6+}, AuF^{5+}, AuF^{4+} and AuF^{3+}. Non-metallated peptide was identified, in the form of $apoF^{6+}$, $apoF^{5+}$ and $apoF^{4+}$. Interestingly auranofin presented some reaction products where the phosphine ligand remained coordinated to Au(I), as exemplified by species such as Et_3PAuF^{6+} (614.64), Et_3PAuF^{5+} (737.37) and Et_3PAuF^{4+} (921.46). $[AuCl(Cy_3P)]^+$ is less reactive than $[AuCl(Et_3P)]^+$ when targeting Sp1 (ZnF3), as unreacted ZnF was still present (Fig. 1.8), with some gold incorporation observed due to the presence of AuF peaks (AuF^{4+} and AuF^{3+}). The reactivity of $[Au(dmap)(Cy_3P)]$ (**5**) is different when compared to the chloride precursor (**4**). No unreacted ZnF is observed, apo-peptide is present ($apoF^{4+}$ and $apoF^{3+}$), and gold is also incorporated (AuF^{4+} at 891.94 and AuF^{3+} at 1188.91 m/z).

1.3.6 Full-Length NCp7 Zinc Finger/SL2 DNA Interaction Inhibition

The macromolecular interaction (binding) between biomolecules can be studied using fluorescence polarization (FP). This technique is based on the tumbling movement of molecules in solution in presence of polarized light. Larger molecules (or interacting systems) will retain the polarization for longer time due to slower tumbling. The binding between full-length NCp7 zinc finger and SL2 DNA was studied by FP. The binding curve is shown in Fig. 1.32. The K_d value was found to be 899 ± 46 nM, using a Michaelis-Menten/Hill model ($n = 1$). The inhibition experiments were set up knowing the maximum binding conditions. Once the binding event was well characterized, inhibition experiments were set up. Figure 1.9 shows the binding behavior of the full-length NCp7 zinc finger with SL2 in the presence of the Cy_3P series, [AuCl(Cy_3P)] (**I-4**) and [Au(dmap)(Cy_3P)]$^+$ (**I-5**), and also auranofin as a representative compound of the Et_3P series. As shown in Fig. 1.9, inhibition was only observed for the Cy_3P series, while auranofin was not able to induce the same binding inhibition, as well as [AuCl(Et_3P)] and [Au(dmap)(Et_3P)]$^+$ (data not shown). The inhibition constants obtained using an exponential trendline for fitting the data obtained for [AuCl(Cy_3P)] and [Au(dmap)(Cy_3P)]$^+$ were of 28.6 and 22.0 μm, respectively. FP demonstrates the final outcome of the auration process, since the natural function of the full-length NCp7 zinc finger relies on its interaction with the viral RNA. SL2 DNA is a representative oligonucleotide for this interaction. The large {(Cy_3P)Au} moiety seems to be required for the inhibition, since Et_3P-containing compounds caused no significant inhibition (auranofin data, as an example, is shown in Fig. 1.9).

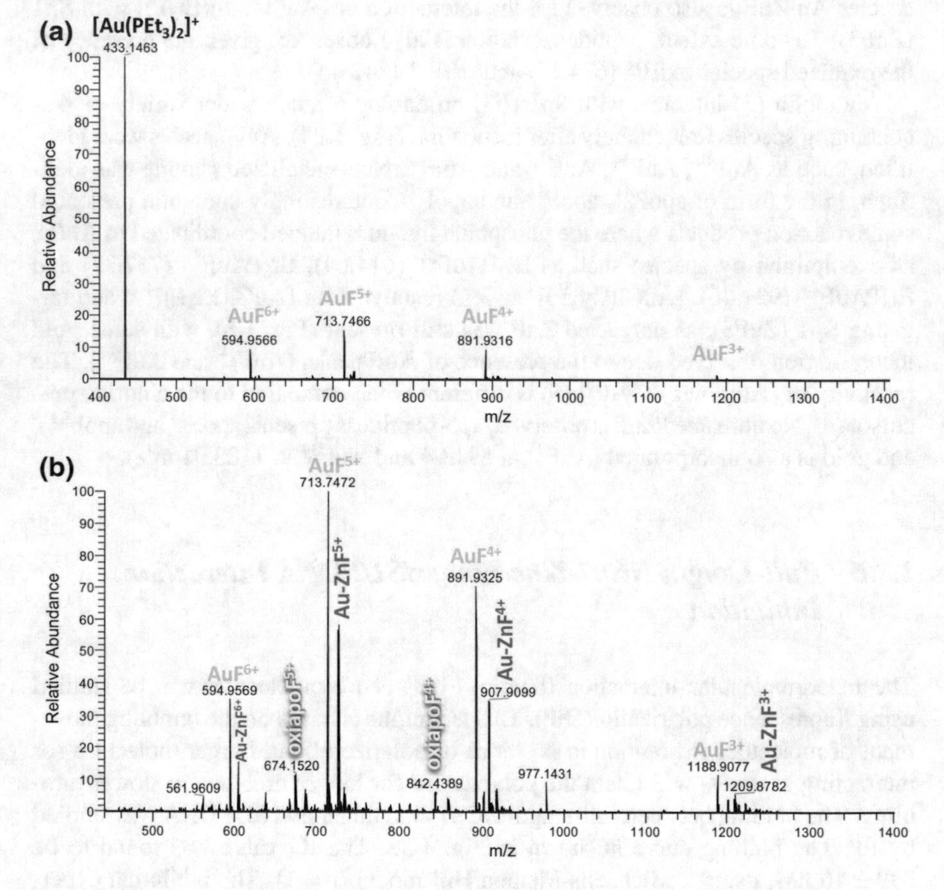

Fig. 1.8 ESI-MS spectrum (positive mode) for the reaction between Sp1 (ZnF3) and **a** [AuCl(Et_3P)]$^+$, **b** [Au(dmap)(Et_3P)]$^+$, **c** [AuCl(Cy_3P)]$^+$ and **d** [Au(dmap)(Cy_3P)]$^+$ immediately upon incubation

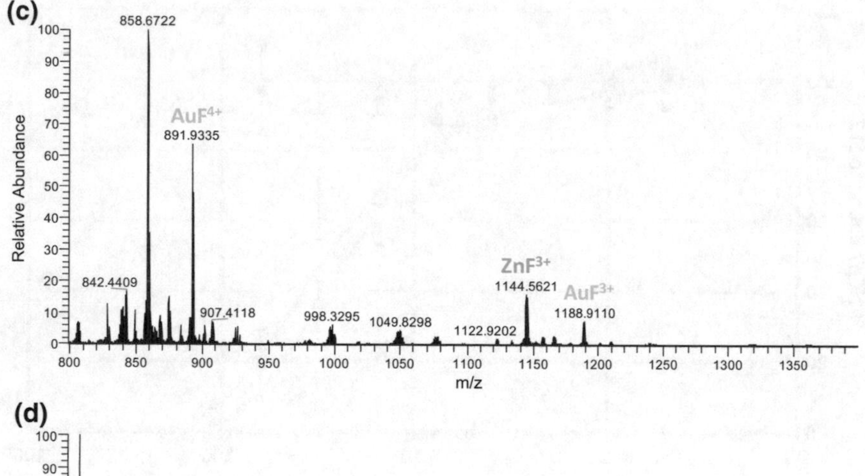

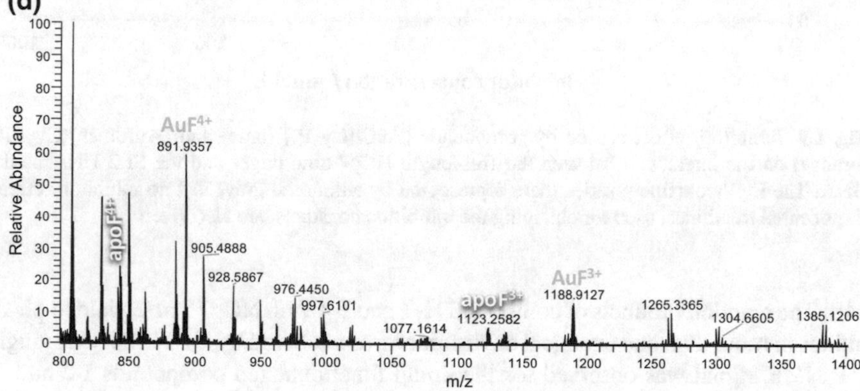

Fig. 1.8 (continued)

1.4 Conclusions

In this chapter we explored structural variations in the phosphine ligand for fine-tuning the reactivity of a series of Au(I) compounds designed for targeting zinc finger proteins. Properties of the phosphine such as steric hindrance and basicity of the phosphorous atom were evaluated. The more basic and less hindered Et_3P as well as the bulkier Cy_3P were selected for this work. The nature of the co-ligand was also evaluated. 4-dimethylaminopyridine (dmap) was selected as an alternative co-ligand based on its π-stacking properties, which can mimic the naturally occurring interaction between NCp7 and nucleic acids, adding an extra specificity property to the designed compounds. Auranofin, the commercially available and widespread Au(I)-phosphine-based metallopharmaceutical was included in this study given the structural similarities with the compounds discussed here.

The replacement of Cl by dmap decreases the reactivity rate of the Au(I) compound dramatically, in terms of the temporal evolution of the aurated species identified by

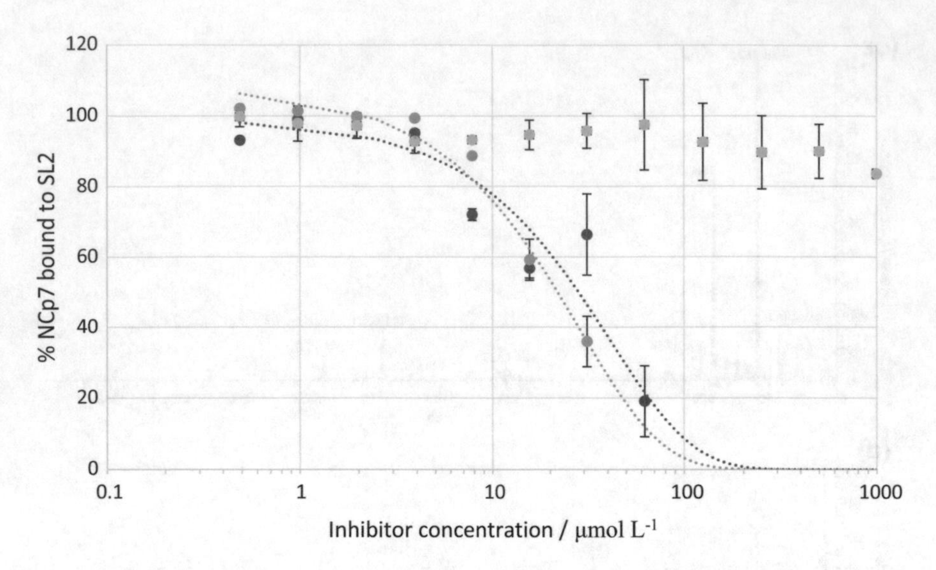

Fig. 1.9 Inhibitory effect caused by compounds [AuCl(Cy_3P)] (blue) and [Au(dmap)(Cy_3P)]$^+$ (orange) on the interaction between the full-length NCp7 zinc finger and the SL2 DNA model DNA. The Et_3P-containing series (here represented by auranofin, grey) had no inhibitory effect. Exponential trendlines, used for obtaining the inhibition constants, are also given

MS. The reaction products of compounds **I-1** and **I-4** with NCp7 ZnF2 yielded phosphines in more than one chemical environment, as observed by ^{31}P NMR. One single ^{31}P NMR signal was observed for the dmap functionalized compounds **I-2** and **I-5**. When comparing the ancillary phosphines, Et_3P-containing compounds seem to be in general less reactive than Cy_3P-containing compounds. Slower reaction rate is often translated into higher metalation selectivity. Combining the effects of the phosphine and ligand L, [Au(dmap)(Et_3P)]$^+$ (**I-2**) was the least reactive compound in the series when targeting NCp7 ZnF2, as characterized by CD, MS and ^{31}P NMR. Auranofin is unique, as observe by ^{31}P NMR and MS/MS data, as it is the only compound that leads to Au-His species. Scheme 1.1 summarizes the Au(I) incorporation mechanisms observed here.

The interaction between the full-length NCp7 zinc finger with the model SL2 DNA sequence was also targeted. Inhibition was closely associated with the steric hindrance of the phosphine ligands. Cy_3P-containing compounds (**I-4** and **I-5**) were the only ones capable of inhibiting the biomolecular interaction, as evaluated by fluorescence polarization. Although the phosphine can be lost throughout the reaction, as evidenced by MS data, it still plays an important role in the inhibition mechanism. An overall proposed inhibition mechanism caused by Cy_3P-containing compounds is presented in Scheme 1.2.

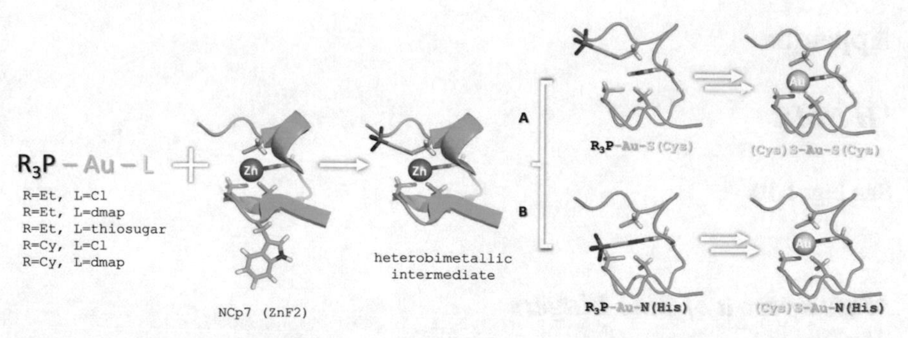

Scheme 1.1 Au(I) incorporation into NCp7 ZnF2 (the same can be extended to the full-length NCp7 zinc finger). The first step is the electrophilic attack of the Au(I)-phosphine compounds on the Zn-coordinated residues, forming a {R_3PAu}-ZnF species. Two alternative pathways open up from this species. The final Au-peptide species depends on both the phosphine and the co-ligand L. **A**. Along with Zn displacement, the {R_3PAu} moiety remains coordinated to a Cys residue and, after loss of the phosphine, a Cys-Au-Cys AuF is obtained. This is the most typical pathway, observed for compounds **I-1**, **I-2**, **I-4** and **I-5**. **b** An alternative pathway was observed for auranofin, where the {R_3PAu} moiety coordinates to a His residue. The final hypothetical AuF obtained from this pathway would have a Cys-Au-His coordination sphere

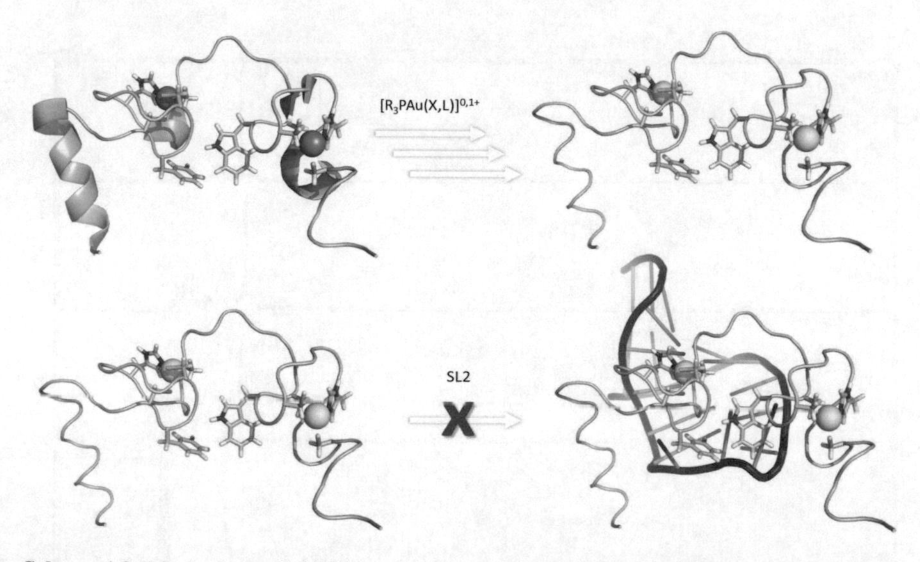

Scheme 1.2 Mechanism of inhibition of the recognition of SL2 model DNA by the full-length NCp7 ZnF caused by the Au(I)-phosphines compounds. Zinc displacement followed by Au(I) replacement leads to loss of secondary structure of the protein. The Au(I)-containing protein adducts caused by the series containing Cy_3P are unable to recognize the model SL2 DNA

Appendix

¹H NMR

See Fig. 1.10.

Mass Spectra of Zinc Fingers

See Figs. 1.11, 1.12 and 1.13.

Ligand Scrambling

See Figs. 1.14, 1.15, 1.16 and 1.17.

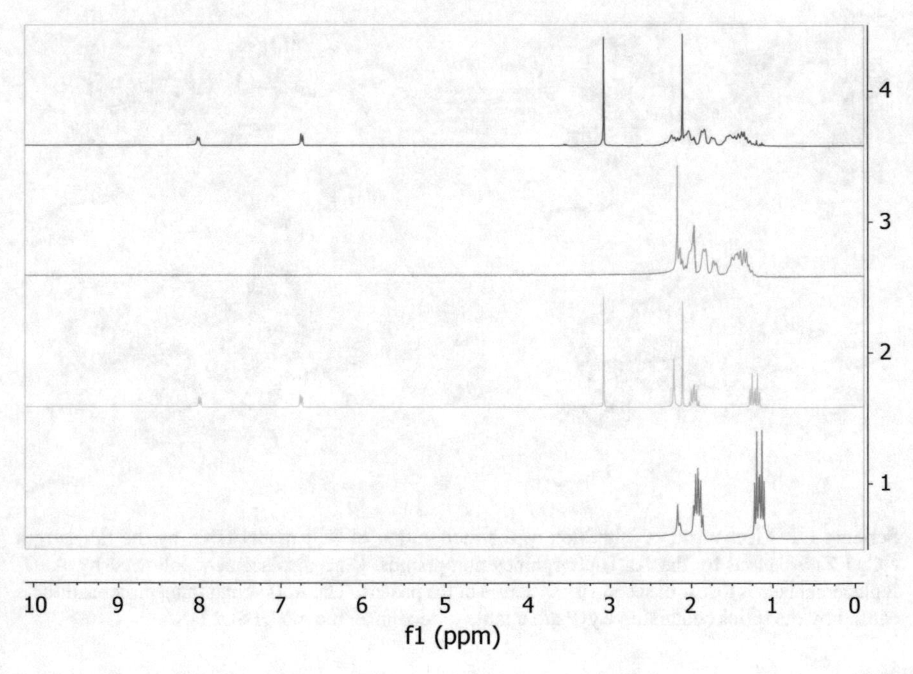

Fig. 1.10 ^{1}H NMR spectra of (**1**) [AuCl(Et$_3$P)], (**2**) [Au(dmap)(Et$_3$P)]$^+$, (**3**) [AuCl(Cy$_3$P)] and (**4**) [Au(dmap)(Cy$_3$P)]$^+$ in deuterated acetonitrile

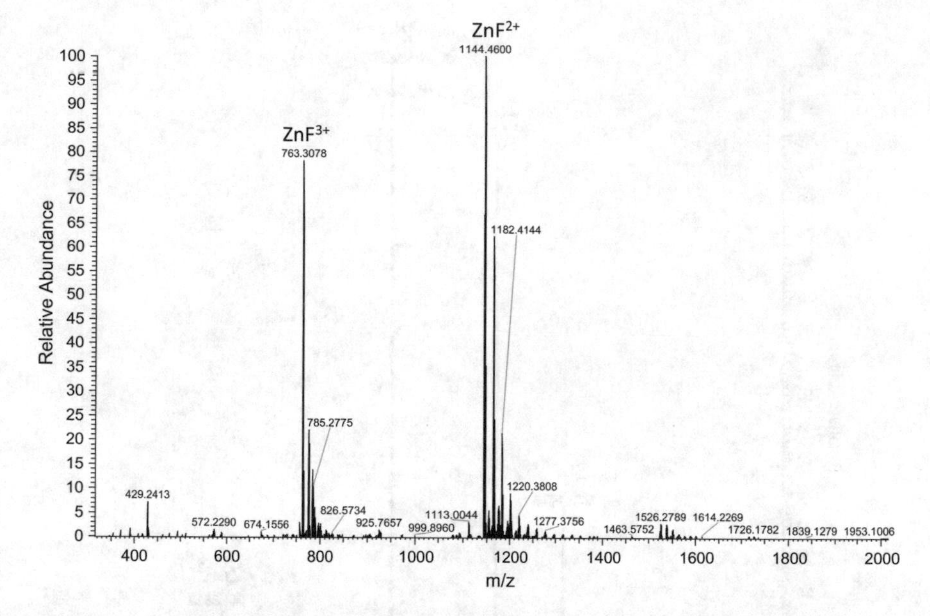

Fig. 1.11 Representative mass spectrum of the HIV-1 nucleocapsid protein NCp7 (ZnF2)

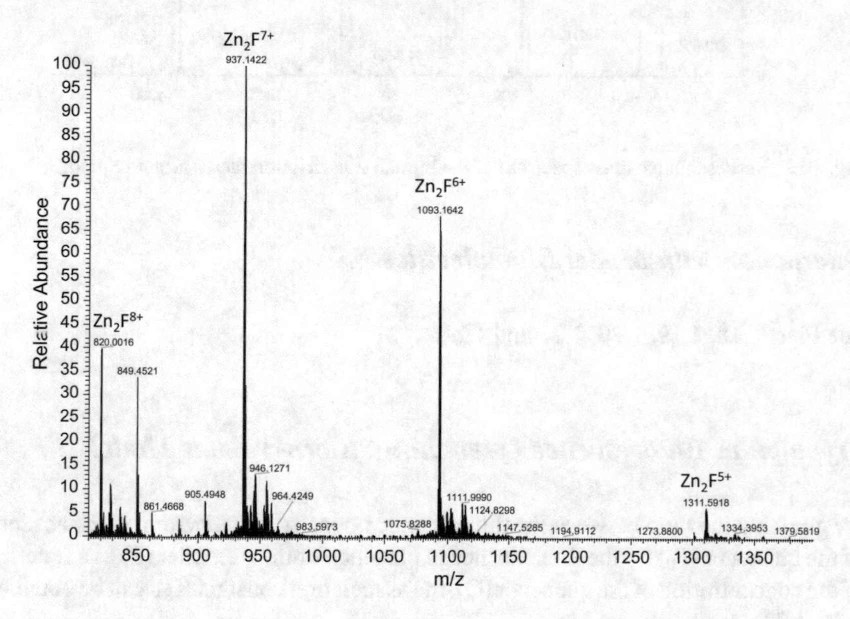

Fig. 1.12 Representative mass spectrum of the HIV-1 nucleocapsid protein NCp7 "full" zinc finger

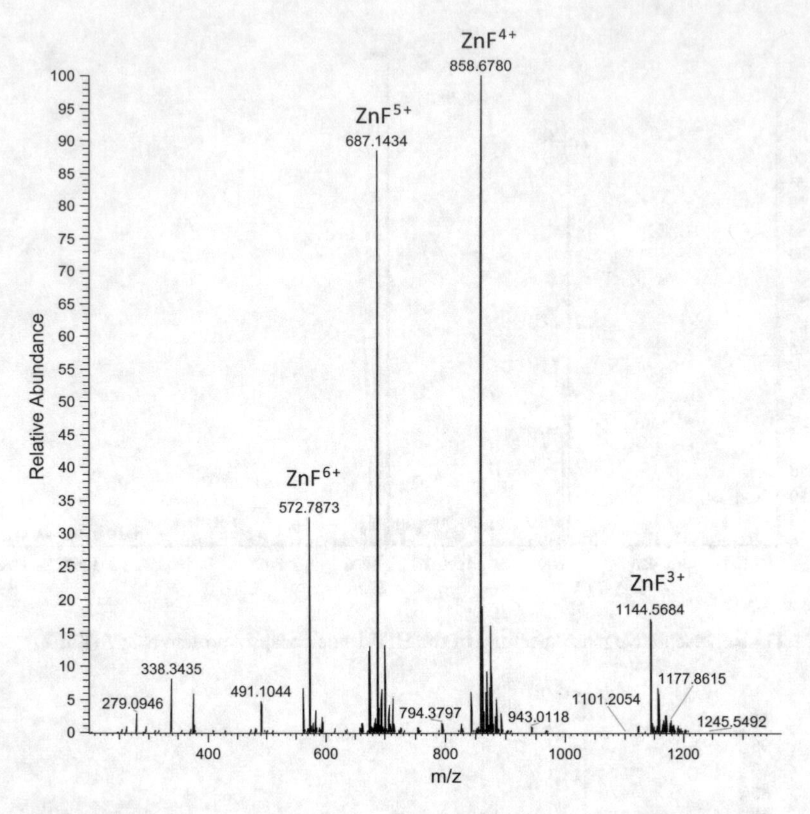

Fig. 1.13 Representative mass spectrum of the human transcription factor protein Sp1 (F3)

Interaction with Model Biomolecules

See Figs. 1.18, 1.19, 1.20, 1.21 and 1.22.

Tryptophan Fluorescence Quenching: Stern-Volmer Model

By graphing the ratio between the fluorescence intensity of the system in the absence of the quencher (F_0) by the fluorescence in presence of the quencher (F) as a function of the concentration of the quencher [Q], the association constant K_{SV} can be obtained (Fig. 1.23).

$$\frac{F_0}{F} = 1 + K_{SV}[Q]$$

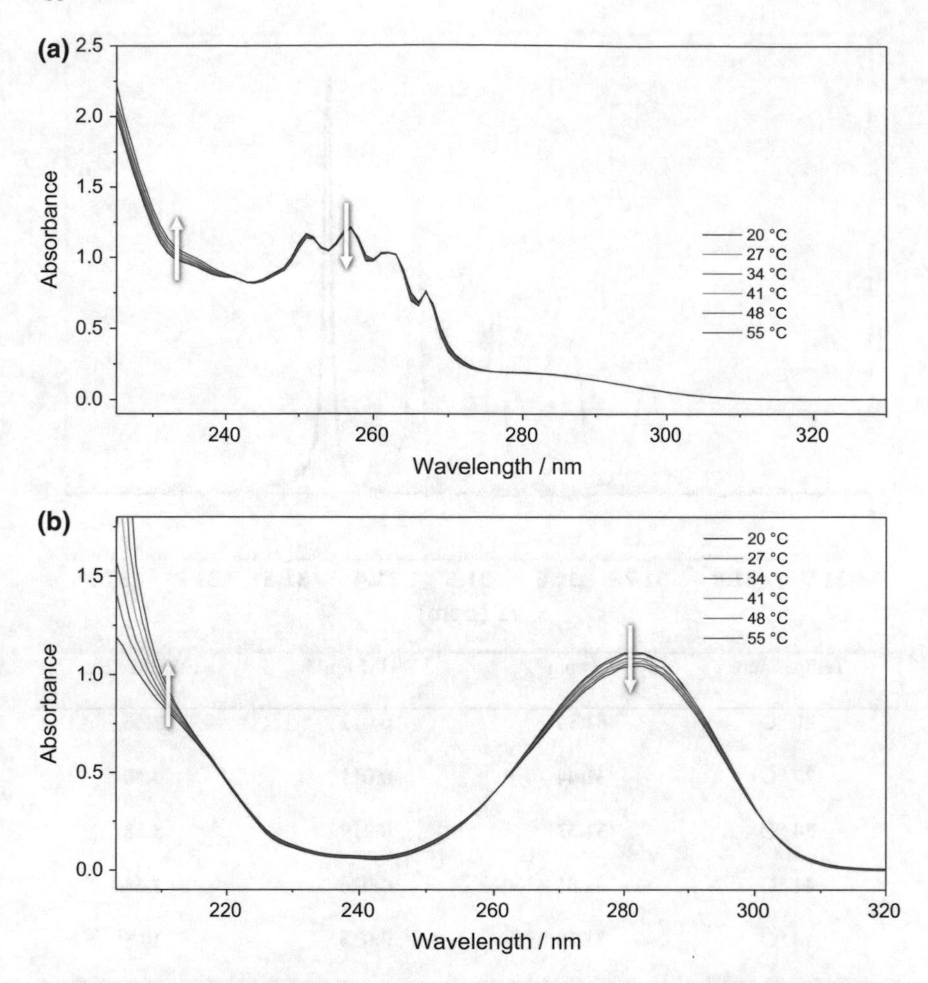

Fig. 1.14 Temperature dependent UV-Vis data for [AuCl(Cy_3P)] (**a**) and [Au(dmap)(Cy_3P)]$^+$ (**b**). The arrows indicate intensity changes as consequence of temperature increase

Interaction with NCp7 ZnF2

See Figs. 1.24, 1.25, 1.26, 1.27, 1.28, 1.29 and 1.30.

Interaction with Sp1 (ZnF3)

See Fig. 1.31.

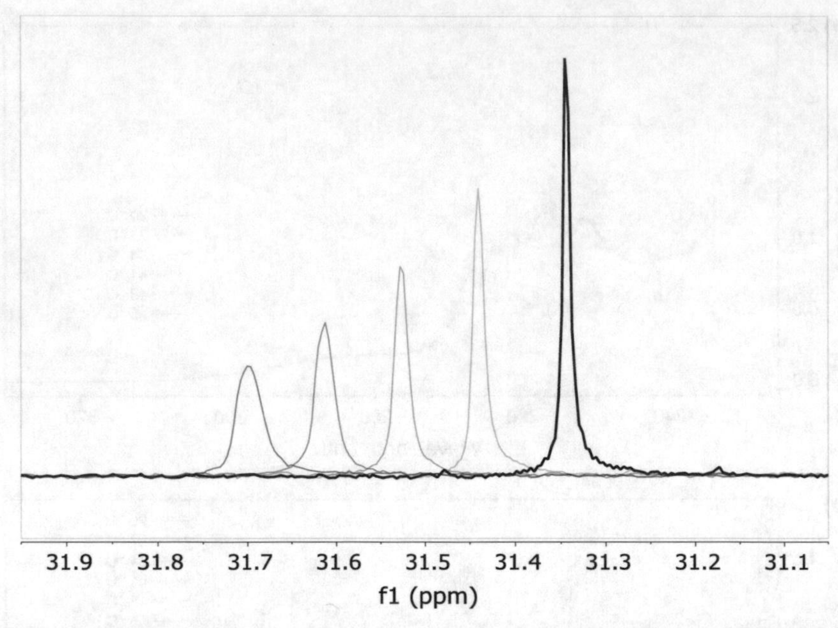

Temperature	δ / ppm	FWHM / ppm	FWHM / Hz
20 °C	31.34	0.012	3.66
27 °C	31.44	0.011	3.40
34 °C	31.53	0.019	5.58
41 °C	31.61	0.026	7.68
48 °C	31.70	0.035	10.6

Fig. 1.15 Temperature effect on the ^{31}P NMR signal of [AuCl(Et_3P)]. Spectral parameters are given for each temperature evaluated. The spectrum acquired at the starting temperature (27 °C) is presented in bold, with the ^{31}P signal getting further deshielded with temperature increase

Full-length NCp7 ZnF/SL2 DNA Interaction

See Fig. 1.32.

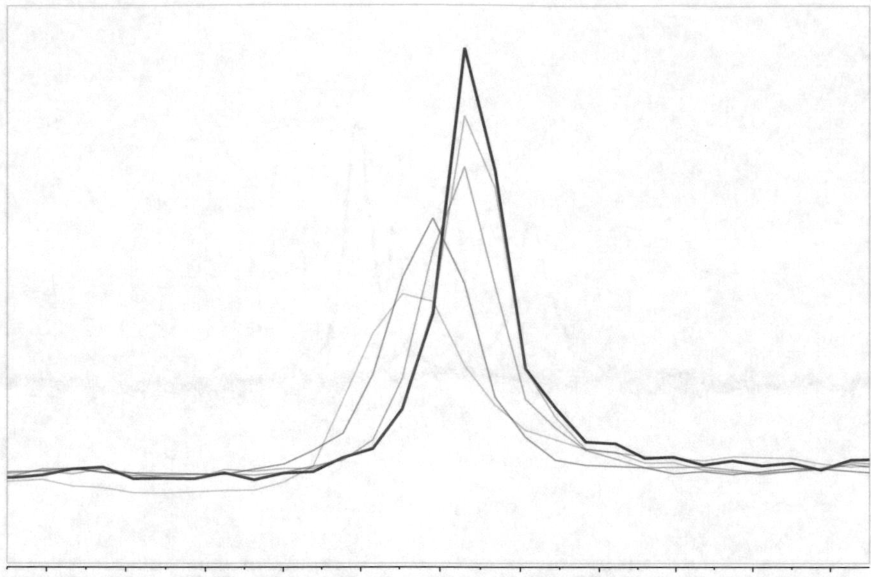

Temperature	δ / ppm	FWHM / ppm	FWHM / Hz
20 °C	53.95	0.009	2.63
27 °C	53.95	0.010	2.94
34 °C	53.95	0.011	3.36
41 °C	53.95	0.013	3.84
48 °C	53.96	0.017	5.24

Fig. 1.16 Temperature dependence on the ^{31}P NMR signal of [AuCl(Cy_3P)]. Spectral parameters are given for each temperature evaluated. The spectrum acquired at the starting temperature (27 °C) is presented in bold, with the ^{31}P signal getting further deshielded with temperature increase

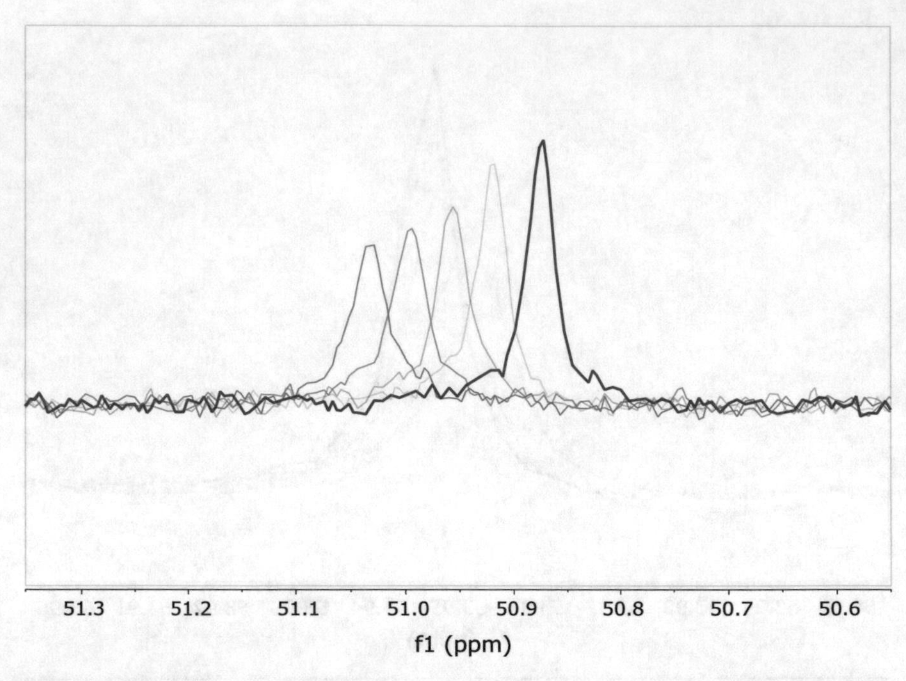

Temperature	δ / ppm	FWHM / ppm	FWHM / Hz
20 °C	50.57	0.031	9.37
27 °C	50.92	0.035	10.5
34 °C	50.96	0.037	11.0
41 °C	50.99	0.041	12.4
48 °C	51.03	0.042	12.7

Fig. 1.17 Temperature dependence on the ^{31}P NMR signal of $[Au(dmap)(Cy_3P)]^+$. Spectral parameters are given for each temperature evaluated. The spectrum acquired at the starting temperature (27 °C) is presented in bold, with the ^{31}P signal getting further deshielded with temperature increase

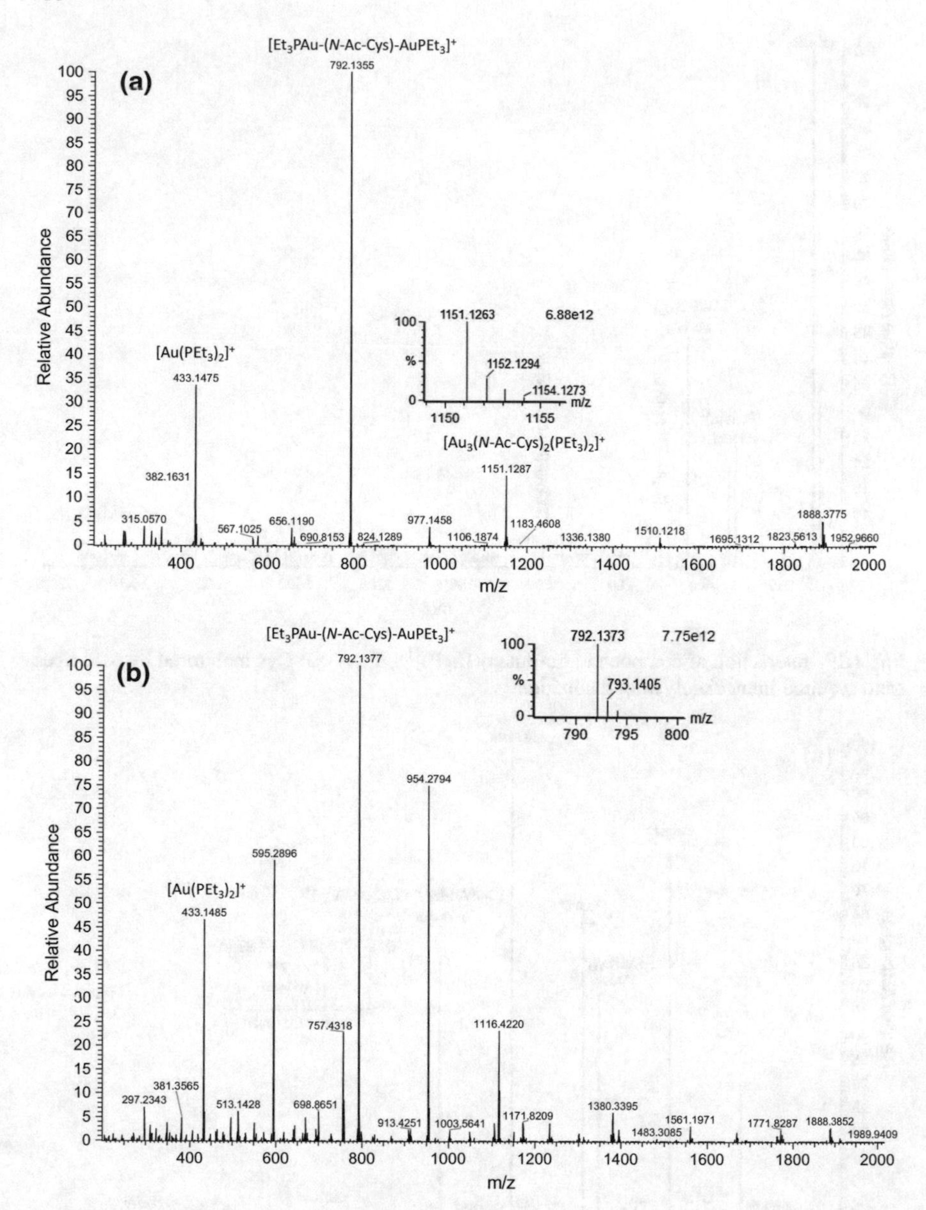

Fig. 1.18 Interaction of compound [AuCl(Et_3P)] with *N*-Ac-Cys monitored by MS. Spectrum acquired **a** immediately after incubation and **b** after 24 h

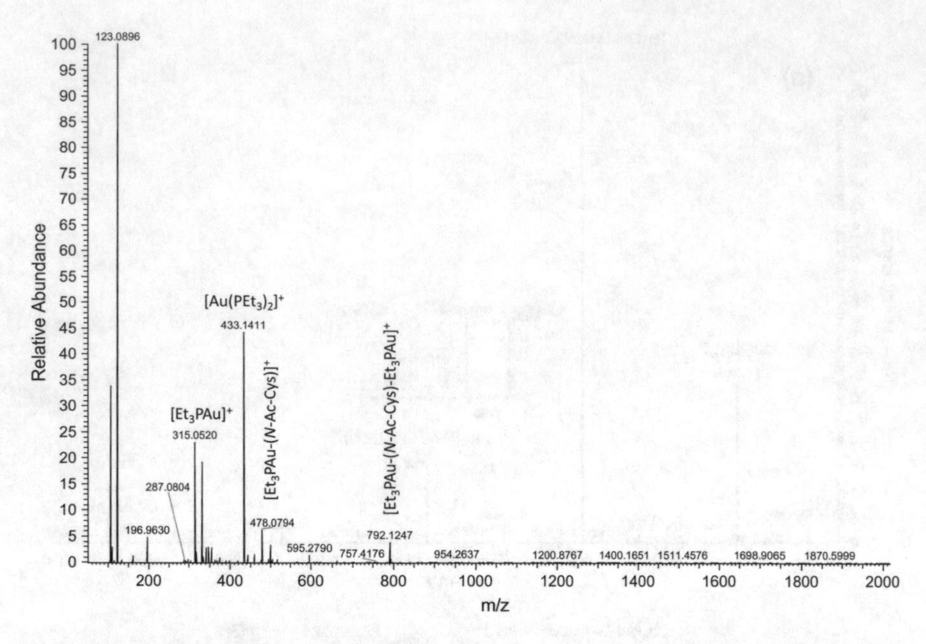

Fig. 1.19 Interaction of compound $[Au(dmap)(Et_3P)]^+$ with *N*-Ac-Cys monitored by MS. Spectrum acquired immediately after incubation

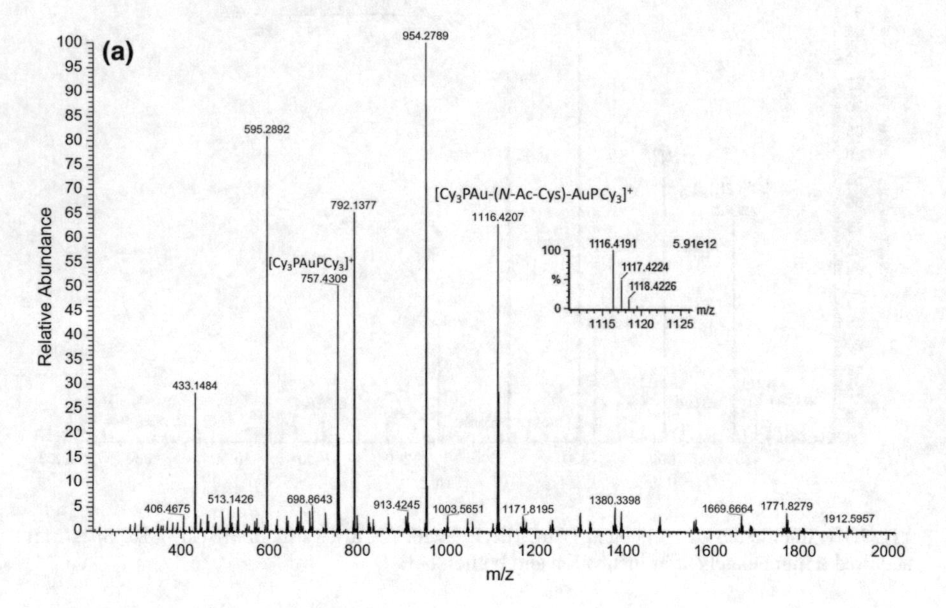

Fig. 1.20 Interaction of compound $[AuCl(Cy_3P)]$ with *N*-Ac-Cys followed by MS. Spectrum acquired **a** immediately after incubation and **b** after 24 h

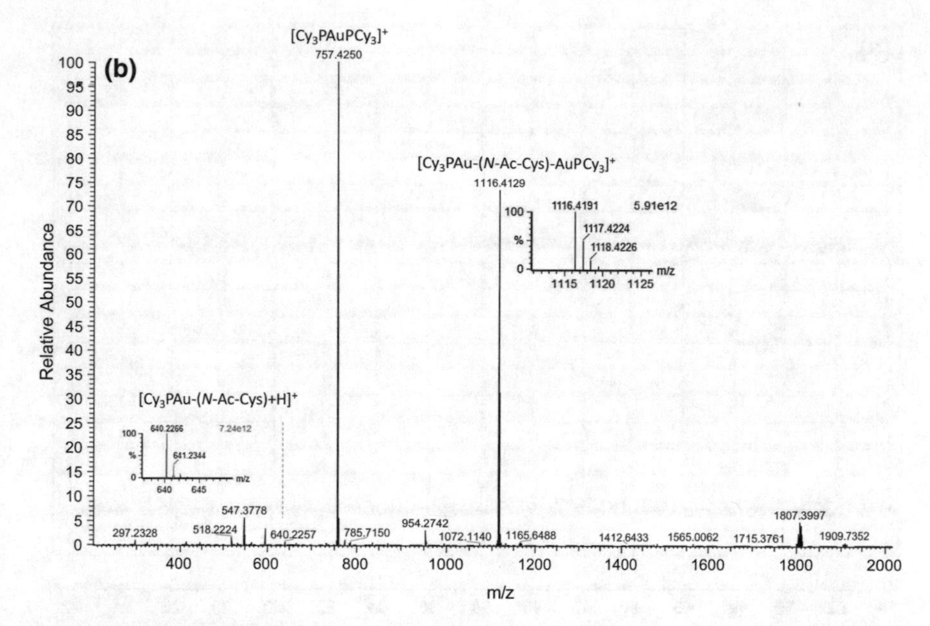

Fig. 1.20 (continued)

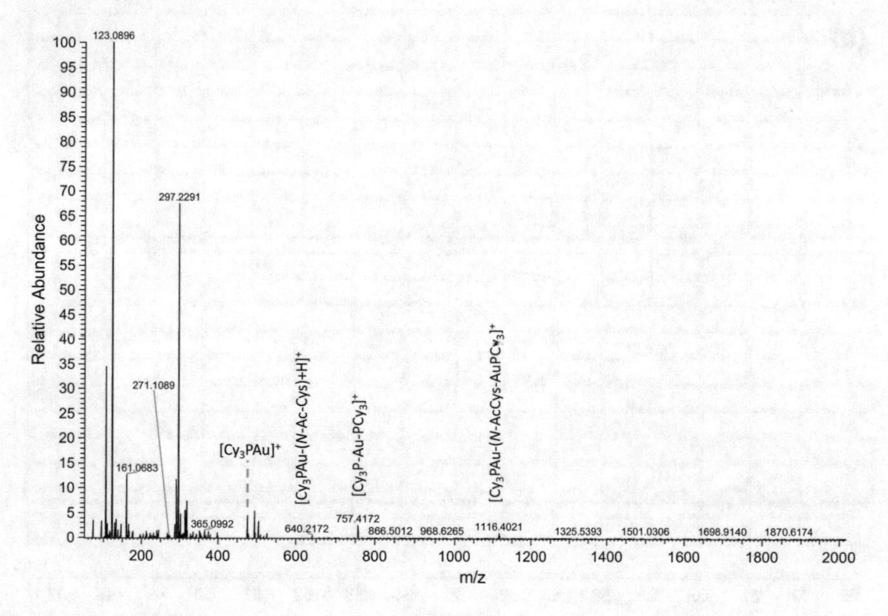

Fig. 1.21 Interaction of compound $[Au(dmap)(Cy_3P)]^+$ with *N*-Ac-Cys monitored by MS. Spectrum acquired immediately after incubation

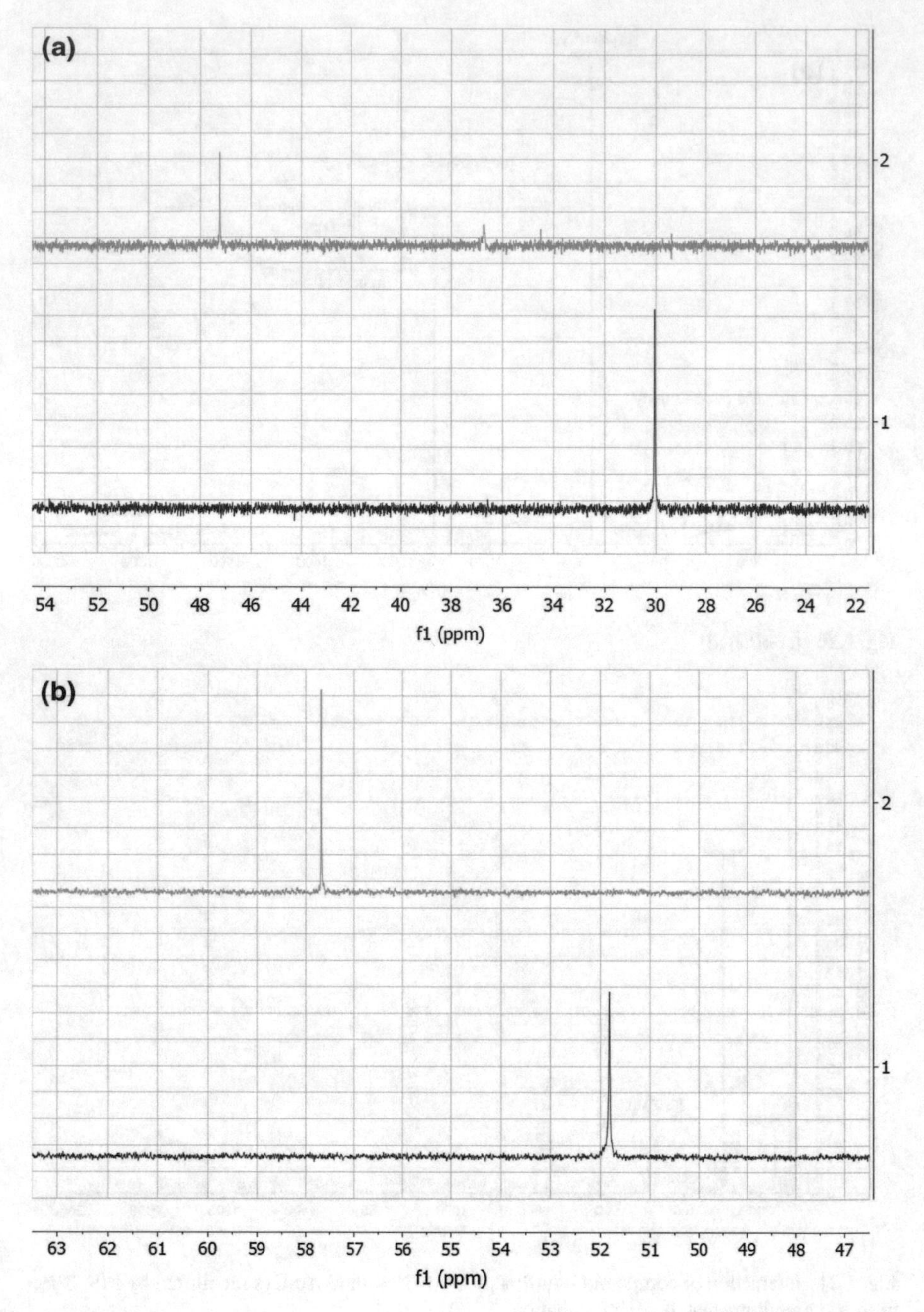

Fig. 1.22 ^{31}P NMR spectra of the reaction products of **a** $[Au(dmap)(Et_3P)]^+$ and **b** $[Au(dmap)(Cy_3P)]^+$ with N-Ac-Cys (**1**) immediately after mixing and (**2**) 24 h after incubation

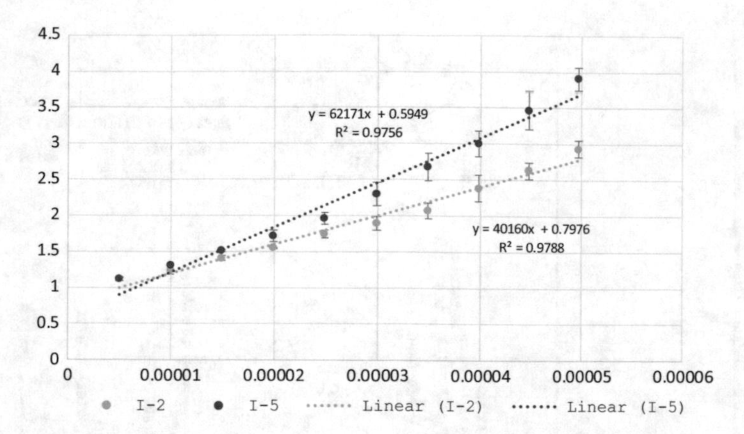

Fig. 1.23 Tryptophan fluorescence quenching assay. Data presented for the dmap-functionalized compounds $[Au(dmap)(Et_3P)]^+$ (**I-2**) and $[Au(dmap)(Cy_3P)]^+$ (**I-5**). Data was treated using the Stern-Volmer model

Interaction with NCp7 ZnF2

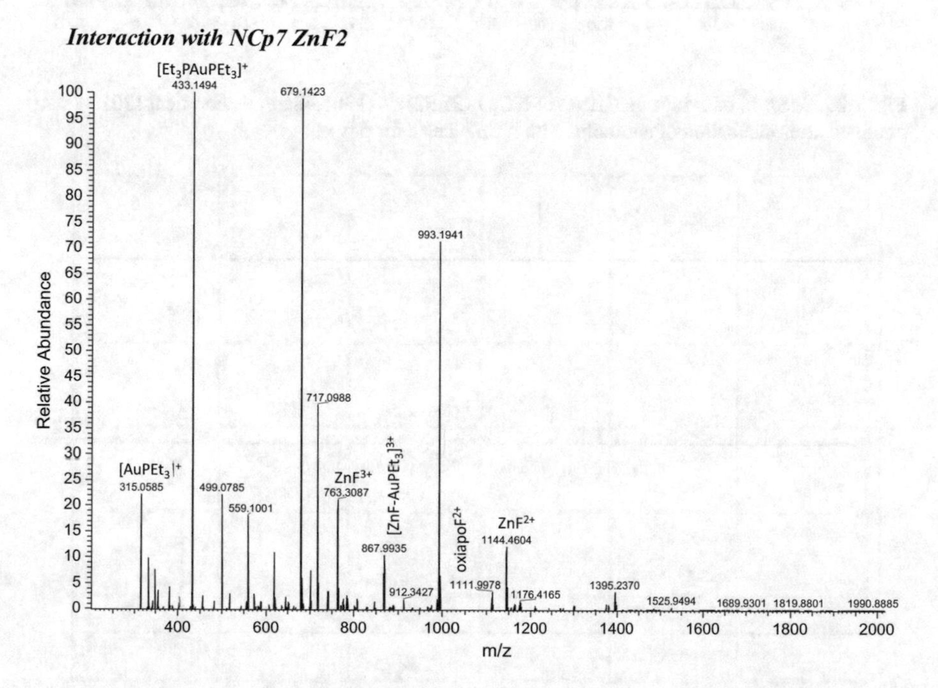

Fig. 1.24 ESI-MS spectrum (positive mode) of the reaction between auranofin and NCp7 ZnF2 obtained after 2 h of incubation

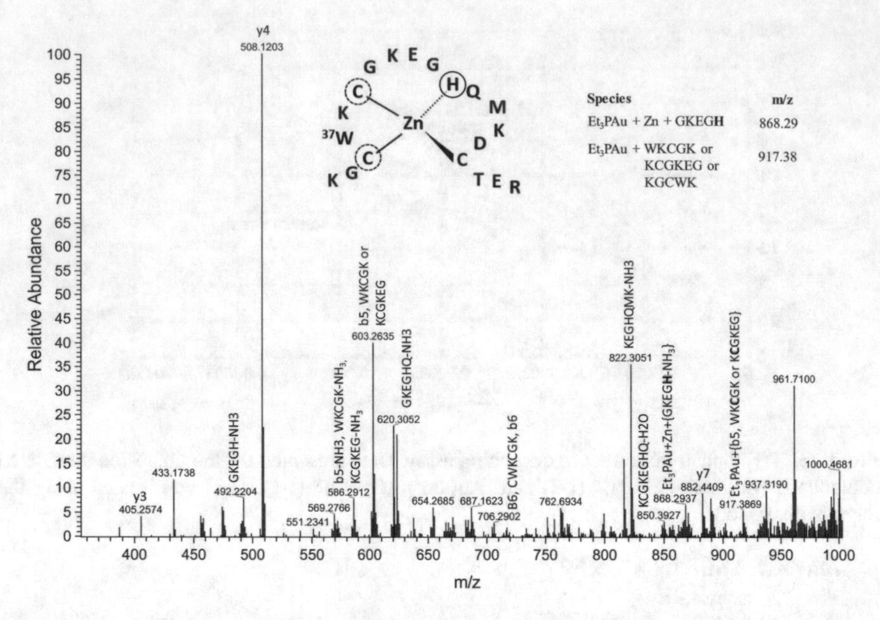

Fig. 1.25 MS/MS of the species $[Et_3PAu\text{-}NCp7\ (ZnF2)]^{2+}$ (1301.48 m/z; theoretical 1301.48 m/z) obtained after incubation of auranofin with NCp7 ZnF2 for 6 h

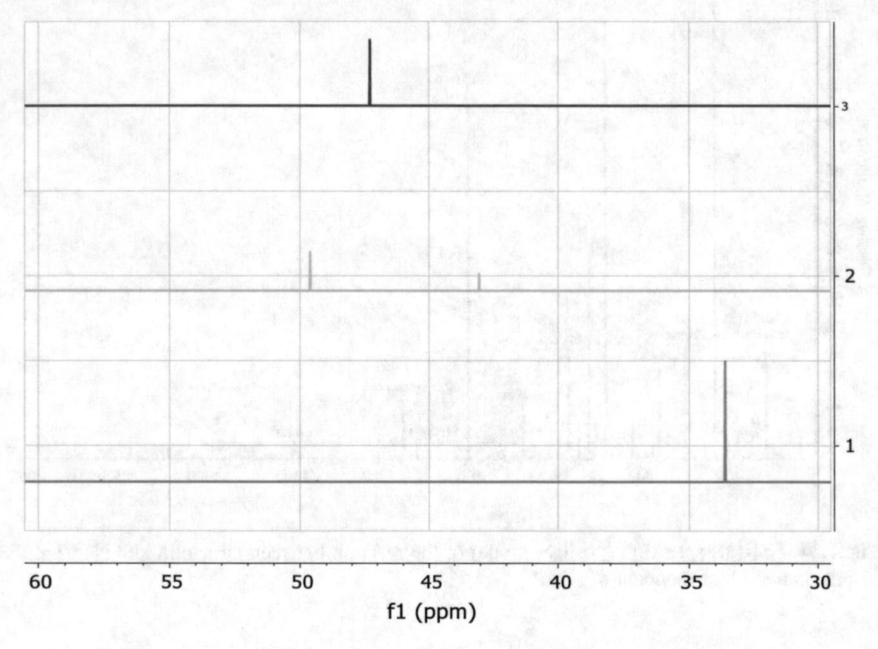

Fig. 1.26 ^{31}P NMR spectra of the reaction product obtained after incubation of NCp7 (ZnF2) with $[AuCl(Et_3P)]$ (**2**) for 6 h and (**3**) after incubation with the full-length NCp7 zinc finger for 8 days. (**1**) For comparison, the ^{31}P NMR spectrum of the free $[AuCl(Et_3P)]$ is also shown

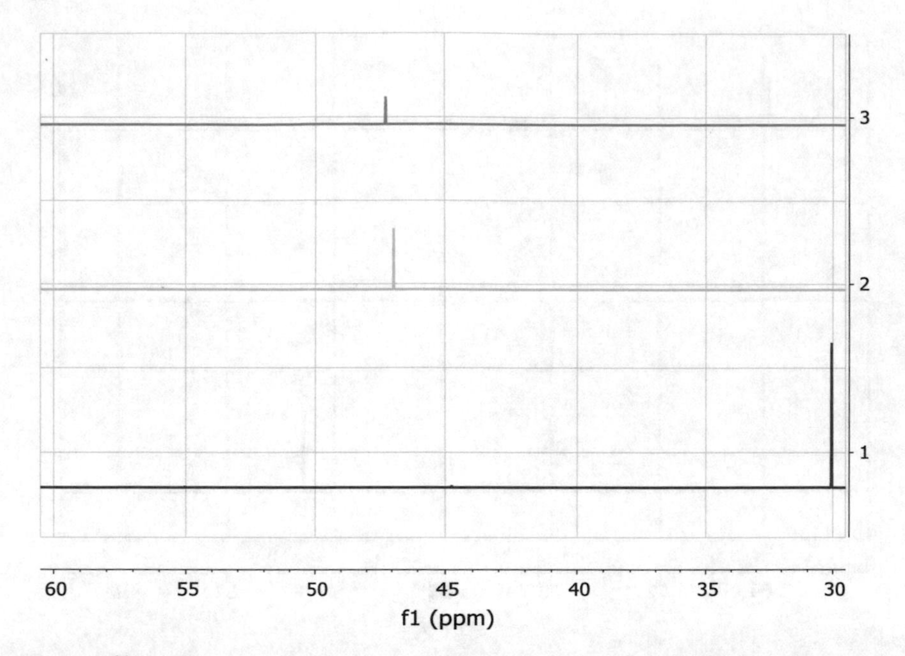

Fig. 1.27 ^{31}P NMR spectra of the reaction product obtained after incubation of NCp7 (ZnF2) with $[Au(dmap)(Et_3P)]^+$ (**2**) for 6 h and (**3**) after incubation with the full-length NCp7 zinc finger for 8 days. (**1**) For comparison, the ^{31}P NMR spectrum of the free $[Au(dmap)(Et_3P)]^+$ is also shown

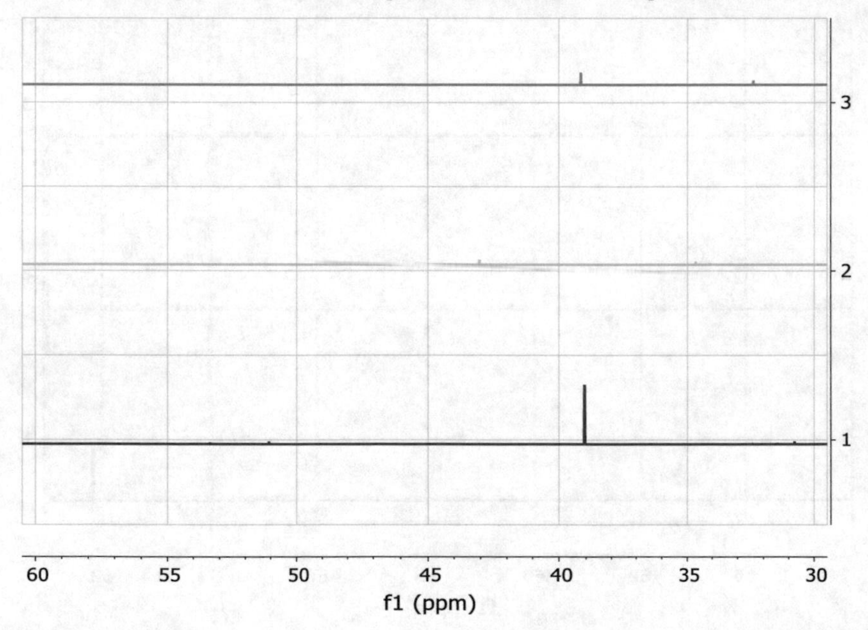

Fig. 1.28 ^{31}P NMR spectra of the reaction product obtained after incubation of NCp7 (ZnF2) with auranofin (**2**) for 6 h and (**3**) after incubation with the full-length NCp7 zinc finger for 8 days. (**1**) For comparison, the ^{31}P NMR spectrum of the free auranofin is also shown

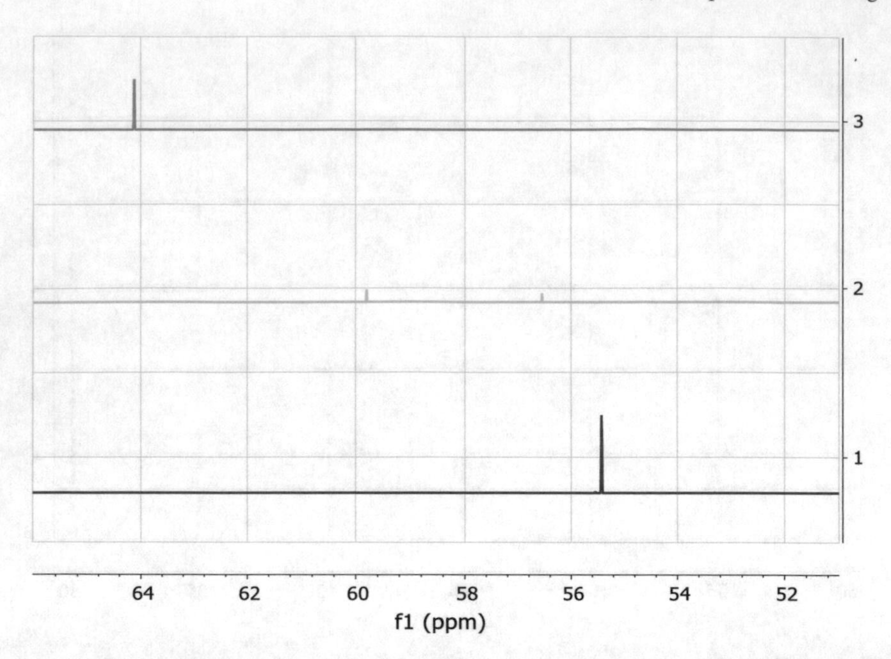

Fig. 1.29 ^{31}P NMR spectra of the reaction product obtained after incubation of NCp7 (ZnF2) with ([AuCl(Cy_3P)]) (**2**) for 6 h and (**3**) after incubation with the full-length NCp7 zinc finger for 8 days. (**1**) For comparison, the ^{31}P NMR spectrum of the free [AuCl(Cy_3P)] is also shown

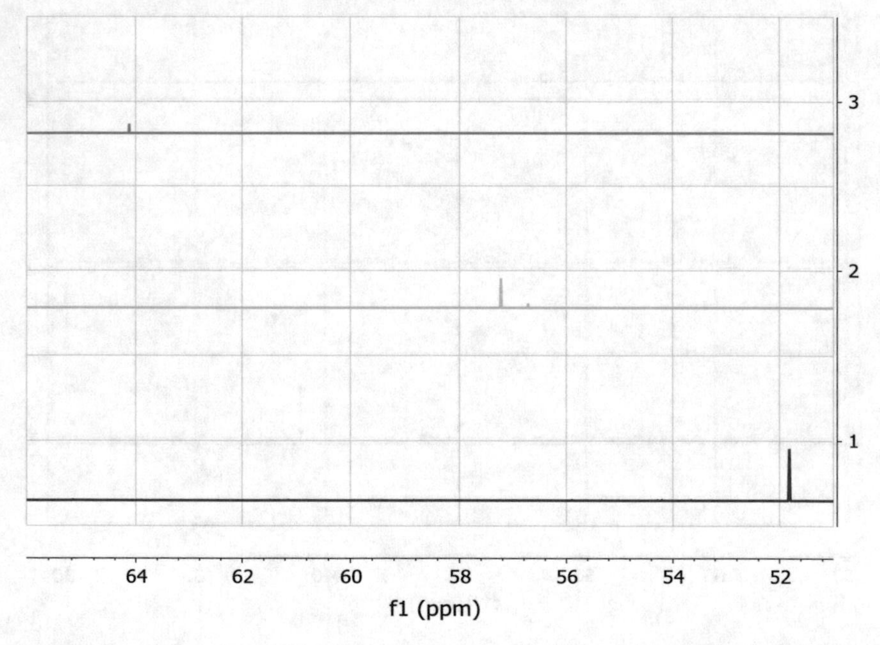

Fig. 1.30 ^{31}P NMR spectra of the reaction product obtained after incubation of NCp7 (ZnF2) with [Au(dmap)(Cy_3P)] (**2**) for 6 h and (**3**) after incubation with the full-length NCp7 zinc finger for 8 days. (**1**) For comparison, the ^{31}P NMR spectrum of the free [Au(dmap)(Cy_3P)] is also shown

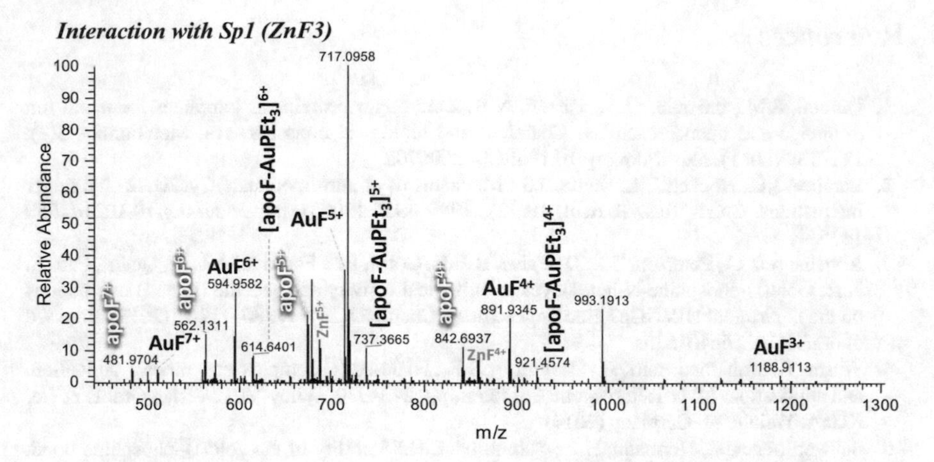

Fig. 1.31 ESI-MS spectrum (positive mode) for the reaction between Sp1 (ZnF3) and auranofin immediately upon incubation

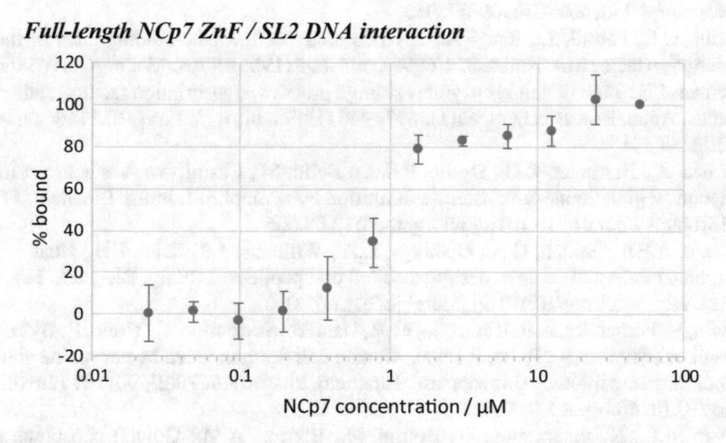

Fig. 1.32 Fluorescence polarization-based binding assay of full-length NCp7 ZnF with the model SL2 DNA sequence

References

1. Quintal, S.M., dePaula, Q.A., Farrell, N.P.: Zinc finger proteins as templates for metal ion exchange and ligand reactivity. Chemical and biological consequences. Metallomics **3**(2), 121–139 (2011). https://doi.org/10.1039/c0mt00070a
2. Larabee, J.L., Hocker, J.R., Hanas, J.S.: Mechanisms of aurothiomalate-Cys2His2 zinc finger interactions. Chem. Res. Toxicol. **18**(12), 1943–1954 (2005). https://doi.org/10.1021/tx050 1435
3. Abbehausen, C., Peterson, E.J., De Paiva, R.E.F., Corbi, P.P., Formiga, A.L.B., Qu, Y., Farrell, N.P.: Gold(I)-phosphine-N-heterocycles: biological activity and specific (ligand) interactions on the C-terminal HIVNCp7 zinc finger. Inorg. Chem. **52**(19), 11280–11287 (2013). https://d oi.org/10.1021/ic401535s
4. Wurm, T., Mohamed Asiri, A., Hashmi, A.S.K.: NHC-Au(I) complexes: synthesis, activation, and application. In: N-Heterocyclic Carbenes, pp. 243–270. Wiley-VCH Verlag GmbH & Co. KGaA, Weinheim, Germany (2014)
5. Schwerdtfeger, P., Hermann, H.L., Schmidbaur, H.: Stability of the gold(I)–phosphine bond. A comparison with other group 11 elements. Inorg. Chem. **42**(4), 1334–1342 (2003). https://d oi.org/10.1021/ic026098v
6. Klug, A.: The discovery of zinc fingers and their development for practical applications in gene regulation and genome manipulation. Q. Rev. Biophys. **43**(1), 1–21 (2010). https://doi.org/10. 1146/annurev-biochem-010909-095056
7. Diakun, G.P., Fairall, L., Klug, A.: EXAFS study of the zinc-binding sites in the protein transcription factor IIIA. Nature **324**(6098), 698–699 (1986). https://doi.org/10.1038/324698a0
8. Coleman, J.E.: Zinc proteins: enzymes, storage proteins, transcription factors, and replication proteins. Annu. Rev. Biochem. **61**(1), 897–946 (1992). https://doi.org/10.1146/annurev.bi.61. 070192.004341
9. De Luca, A., Hartinger, C.G., Dyson, P.J., Lo Bello, M., Casini, A.: A new target for gold(I) compounds: glutathione-S-transferase inhibition by auranofin. J. Inorg. Biochem. **119**, 38–42 (2013). https://doi.org/10.1016/j.jinorgbio.2012.08.006
10. Peacock, A.F.A., Bullen, G.A., Gethings, L.A., Williams, J.P., Kriel, F.H., Coates, J.: Gold-phosphine binding to de novo designed coiled coil peptides. J. Inorg. Biochem. **117**, 298–305 (2012). https://doi.org/10.1016/j.jinorgbio.2012.05.010
11. Gandin, V., Fernandes, A.P., Rigobello, M.P., Dani, B., Sorrentino, F., Tisato, F., Björnstedt, M., Bindoli, A., Sturaro, A., Rella, R., et al.: Cancer cell death induced by phosphine gold(I) compounds targeting thioredoxin reductase. Biochem. Pharmacol. **79**(2), 90–101 (2010). https://d oi.org/10.1016/j.bcp.2009.07.023
12. Karver, M.R., Krishnamurthy, D., Bottini, N., Barrios, A.M.: Gold(I) phosphine mediated selective inhibition of lymphoid tyrosine phosphatase. J. Inorg. Biochem. **104**(3), 268–273 (2010). https://doi.org/10.1016/j.jinorgbio.2009.12.012
13. Karver, M.R., Krishnamurthy, D., Kulkarni, R.A., Bottini, N., Barrios, A.M.: Identifying potent, selective protein tyrosine phosphatase inhibitors from a library of Au(I) complexes. J. Med. Chem. **52**(21), 6912–6918 (2009). https://doi.org/10.1021/jm901220m
14. Hormann-Arendt, A.L., Shaw, C.F.: Ligand-scrambling reactions of cyano(trialkyl/triarylphosphine) gold(I) complexes: examination of factors influencing the equilibrium constant. Inorg. Chem. **29**(23), 4683–4687 (1990). https://doi.org/10.1021/ic 00348a019
15. Xiao, J., Shaw, C.F.: Phosphorus-31 NMR studies of the formation of a (cysteine-34)(μ-thiolato)bis(gold(I) triethylphosphine) species of bovine serum albumin and a related model titration. Inorg. Chem. **31**(18), 3706–3710 (1992). https://doi.org/10.1021/ic00044a010
16. Corbi, P.P., Quintão, F.A., Ferraresi, D.K.D., Lustri, W.R., Amaral, A.C., Massabni, A.C.: Chemical, spectroscopic characterization, and in vitro antibacterial studies of a new gold(I) complex with N-acetyl-L-cysteine. J. Coord. Chem. **63**(8), 1390–1397 (2010). https://doi.org/ 10.1080/00958971003782608

17. Castiglione Morelli, M.A., Ostuni, A., Matassi, G., Minichino, C., Flagiello, A., Pucci, P., Bavoso, A.: Spectroscopic investigation of auranofin binding to zinc finger HIV-2 nucleocapsid peptides. Inorganica Chim. Acta **453**, 330–338 (2016). https://doi.org/10.1016/j.ica.2016.08.012
18. Laskay, Ü.A., Garino, C., Tsybin, Y.O., Salassa, L., Casini, A., Laskay, U.A., Garino, C., Tsybin, Y.O., Salassa, L., Casini, A.: Gold finger formation studied by high-resolution mass spectrometry and in silico methods. Chem. Commun. **51**(9), 1612–1615 (2015). https://doi.org/10.1039/C4CC07490D
19. Tolman, C.A.: Steric effects of phosphorus ligands in organometallic chemistry and homogeneous catalysis. Chem. Rev. **77**(3), 313–348 (1977). https://doi.org/10.1021/cr60307a002

Chapter 2
Probing the Protein: Ion Mobility Spectrometry

2.1 Introduction

The development of soft ionization techniques such as ESI and MALDI turned mass spectrometry into one of the most important tools for studying the interaction of metallodrugs with complex biological targets and also for identifying and characterising the interactions happening on the binding sites [1]. The typical structural techniques, based on X-ray crystallography or NMR, assume one preferred binding site for the metal ion, although in many cases multiple binding possibilities can exist. An example of this diversity is the displacement of Zn(II) in zinc finger peptides by Au(I) and Au(III) compounds, leading to the formation of "gold fingers". These metal replacements have interesting biological consequences, directly derived from the differences in coordination geometry between the intrinsic tetrahedral Zn(II) and the exogenous linear Au(I) and square-planar Au(III), with potential therapeutic applications through inhibition of nucleic acid regnition of the parent zinc fingers [2].

Mass spectrometry has identified multiple $\{Au_nF\}$ ($n = 1, 2, 3$) species following Zn(II) displacement in both the C-terminal finger of the HIV NCp7 nucleocapsid protein, and the finger 3 of Sp1 transcription factor for a series of (dien)Au(III)(N-heterocycle) compounds, a feature easily understood by the presence of multiple cysteines in the zinc finger template [3]. Similar results have been observed using poly(adenosine diphosphate(ADP)-ribose) polymerase 1 (PARP-1) [4]. Zn(II) ions from the full-length NCp7 zinc finger are also displaced by Au(dien)(9-ethylguanine)]$^{3+}$, producing $\{Au_nF\}$. This gold complex shows micromolar efficacy in inhibiting viral infectivity, consistent with the ability to alter the native structure [5]. Auration of zinc finger peptides represents a topic of wide interest in bioinorganic chemistry as well as a fundamental problem for inorganic chemistry, since we are evaluating the replacement of a tetrahedrally-coordinated Zn(II) by a linear-

---

Parts of this chapter has been reproduced with permission from Wiley. https://onlinelibrary.wiley.com/doi/pdf/10.1002/anie.201612494.

R. E. Ferraz de Paiva, *Gold(I,III) Complexes Designed for Selective Targeting and Inhibition of Zinc Finger Proteins*, Springer Theses,
https://doi.org/10.1007/978-3-030-00853-6_2

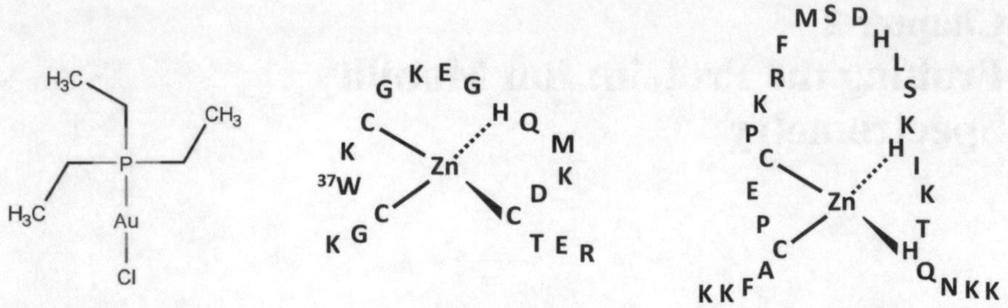

I-1 NCp7 (ZnF2) Sp1 (ZnF3)

Fig. 2.1 The interaction of compound [AuCl(Et$_3$P)] with the she selected zinc finger proteins NCp7 (ZnF2) and Sp1 (ZnF3) was investigated by TWIM-MS coupled with collision induced dissociation (CID)

coordinated Au(I). The compound [AuCl(Et$_3$P)] (**I-1**) was selected as Au(I) donor, and the interaction with NCp7 (ZnF2) and Sp1 (F3) were studied by traveling-wave ion mobility mass spectrometry (TWIM-MS) coupled with collision induced dissociation (CID). The structure of the model compound and that of the selected ZnFs are shown in Fig. 2.1. This approach allowed inequivocal elucidation of specific binding sites and modes of gold-modified NCp7 (ZnF2) and Sp1 (ZnF3). While linear Cys-Au-Cys is indicated for NCp7 (ZnF2), a Cys-Au-His mode is indicated for the Sp1 (ZnF3) case.

2.2 Experimental Section

2.2.1 Materials

The complex [AuCl(Et$_3$P)] was synthesized following the same procedure reported on Part I—Chap. 1 [6]. The peptides NCp7 (ZnF2) and Sp1 (ZnF3) were purchased from GenScript. Protein sequences:

```
                     10         20
NCp7 ZnF2    KGCWKCGKEG HQMKNCTER
Sp1 ZnF3     KKFACPECPK RFMSDHLSKH IKTHQNKK
```

2.2.2 Preparation of Zinc Finger and Gold Finger

The preparation of zinc finger followed published methods [7]. The apopeptide was dissolved in deionized water at a concentration of 1 mmol L^{-1}. Zinc acetate

(1.2:1 mol/mol per zinc binding domain) was added to the solution and the pH was adjusted to 7.0 using NH_4OH. The zinc finger solution was incubated for 2 h at 37 °C before any experiment. Secondary structure characterization of the zinc finger (ZnF) was monitored using ESI-MS and CD spectroscopy. Gold finger peptides were prepared by mixing [AuCl(Et_3P)] in methanol with an equimolar amount of zinc finger peptide at 20 μmol L^{-1} concentration at room temperature. The reaction time allowed before injection was 2 h for the NCp7 peptide whereas the Sp1 peptide reacted immediately. The mixture was then diluted with water/methanol 50% v/v to yield a concentration of 0.2 μmol L^{-1} for electrospray ionization.

2.2.3 *Travelling-Wave Ion Mobility Mass Spectrometry (TWIM-MS)*

Waters Synapt G2-Si travelling wave ion mobility instrument was used. Ions were generated in a nanoelectrospray ionization source (nESI) at positive mode and transferred to vacuum. The experimental settings for conformer-selected collision induced dissociation (CID) experiment were: source temperature, 80 °C; capillary voltage, 3 kV; sample cone, 50 V; source offset, 25 V; cone gas flow, 50 L/H; trap collision energy, 4 V; trap gas flow, 2 mL/min; helium cell gas flow, 180 mL/min; Ion Mobility Spectrometry (IMS) gas flow, 90 mL/min; trap DC bias, 45 V; IMS wave height, 40 V; IMS wave velocity, 1500 m/s. Transfer collision energy for NCp7 AuF2 and Sp1 AuF3 is 35 and 45 eV, respectively.

Conformer-Selected fragmentation peaks were analyzed using Protein-Prospector (http://prospector.ucsf.edu/prospector/mshome.htm). Masslynx 4.0 was used for the data processing and isotope modelling. Collision Cross-Section (CCS) Measurements: calibration of experimental arrival times into collision cross-section values was performed using polyalanine by the automated calibration routine and validated by leucine enkephalin.

2.3 Results and Discussion

The thiophilicity and soft Lewis acid electrophile nature of Au(I) makes sulfur-rich biomolecules logical as cellular targets. In this respect the interference with thioredoxin-thioredoxin reductase is a general consensus as a potential mechanism for the biological activities of gold drugs [8–12]. Early studies on Au(I) compounds such as aurothiomalate suggested possible zinc finger inactivation as contributing to their anti-arthritic activity [13, 14]. Using [AuCl(PPh_3)], both {(PPh_3)AuF} and {Au_nF} species are formed with NCp7 ZnF2 and Sp1 ZnF3, where the transcription factor tends to show a higher propensity for the {Au_nF} over the gold phosphine adduct [15–17]. The properties of [AuCl(PR_3)] make it an ideal system for probing cysteine nucleophilicity in biomolecules, directly analogous to "organic" electrophiles such as maleimide and iodoacetamide. The distal Cys49 is indicated as the initial site of attack of N-ethylmaleimide (NEM), with a near complete reduction in reaction rate when the modified full-length NCp7 ZnF was complexed with nucleic acids [15].

We chose to study the products from [AuCl(Et_3P)], structurally similar to auranofin [18], with two zinc finger sequences representing the C-terminal finger of the $Cys_2HisCys$ (Cys_3His) NCp7 (ZnF2) and the Cys_2His_2 motif of Sp1 (F3). In both cases {(Et_3P)AuF} and {AuF} species were formed with higher intensity for the "gold finger" {AuF} species [19]. We chose the {AuF} species at 4+ charge state for a detailed study. The peaks were then isolated for further IMS analysis. The arrival time distributions (ATDs) of these isolated species are shown in Fig. 2.2, along with the control NCp7 (ZnF2). As expected, the "native" NCp7 C-terminal zinc finger peptide existed as a single conformer (Fig. 2.2a). For the gold(I) finger formed from C-terminal NCp7 (ZnF2), three peaks were observed in the ATD indicating the {AuF} adduct exists as three isomeric species (labelled as x, y and z) in the gas phase, with one major conformer dominant (Fig. 2.2b). The ATD for the equivalent {AuF} adduct from the C-terminal Sp1 (F3) is simpler exhibiting only one peak indicating the existence of a single conformer in the gas phase (Fig. 2.2c). The difference in ATD suggests differing reactivity preferences of the gold(I) drug towards cysteine within zinc finger peptides.

To further investigate the nature of these conformational changes upon auration of the zinc finger peptides, we measured the average collision cross-section of the zinc finger and {AuF} peptides of the 4+ charge state, Fig. 2.2. Under the same conditions, the major {AuF} conformer derived from NCp7 ZnF2 appears slightly more compact ($\Omega = 174.53$ Å^2) compared to the zinc finger peptide itself ($\Omega = 179.83$ Å^2), as we can see from the smaller CCS and shorter drift time. On the other hand the Sp1-derived {AuF} has a slightly longer ATD and now a significantly higher CCS of $\Omega =$ 217.19 Å^2, implying a significantly less compact structure.

To understand these differences in CCS and to identify the auration sites of the various {AuF} conformers we selected a narrow arrival time range for each conformer and subjected the peaks to CID fragmentation. The selected regions are represented in Fig. 2.3. Crosslinking is destroyed upon CID fragmentation producing a discrete set of shorter but still gold-bound fragments from which the original binding sites can be deduced, as observed previously in MS/MS studies of a platinum cross-linked peptide [20].

The MS/MS spectra of the major and minor conformers of the {AuF} derived from NCp7 ZnF2 at 4+ charge state are shown in Fig. 2.4. The identified gold-binding fragments from the MS/MS spectrum of the major conformer are shown in Table 2.1.

Analysis of the CID spectrum of the major conformer revealed that Au^+ mainly bound at the Cys36 and Cys49 as we can observe the aurated a_3-NH_3, y_4, y_6, y_7 and y_8 fragments. The observation of aurated y_{11} to y_{15} cannot give a definite assignment of the auration site but it is consistent with the auration of Cys49. Considering the tendency of gold(I) to form two-coordinate, linear complexes with thiolates, it

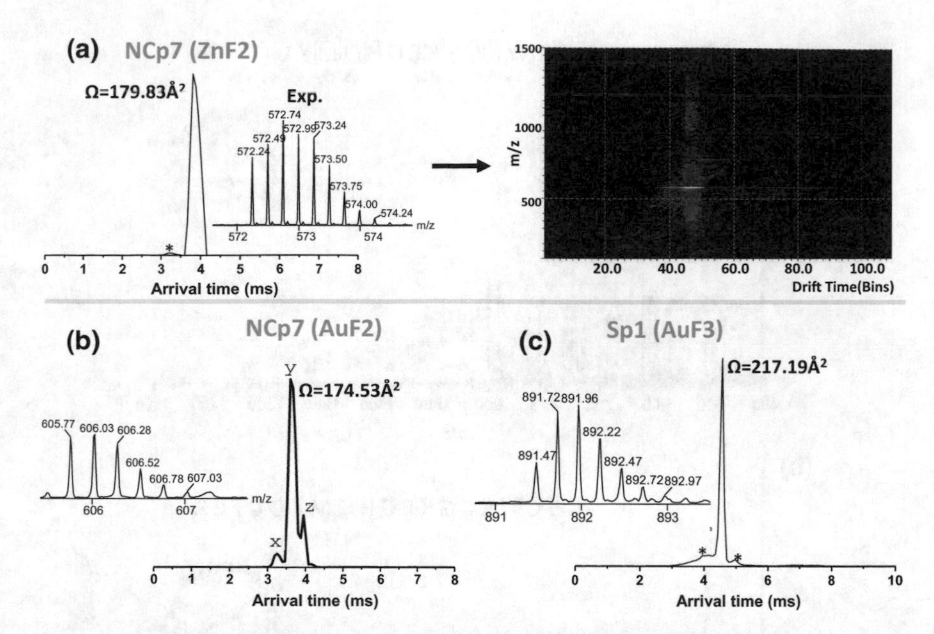

Fig. 2.2 Arrival-time distributions (ATDs) as determined by TWIM-MS, and ESI-MS spectra of the **a** NCp7 (ZnF2), used as control, along with the plot of m/z versus drift time (bins) for the separated NCp7 (ZnF2) at 4+ charge state; **b** NCp7 (AuF2), with three conformers (x, y and z) identified; and **c** Sp1 (AuF3) at 4+ charge state; non-dominant peaks marked with * do not correspond to a AuF conformer

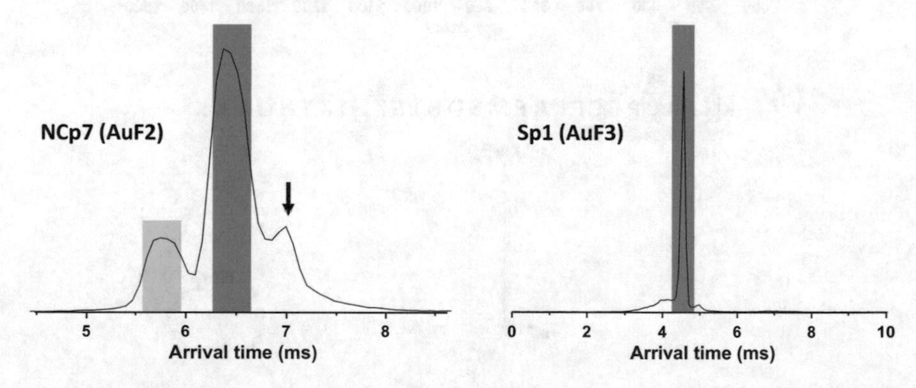

Fig. 2.3 Narrow range of ATDs selected for the CID MS/MS experiments on the NCp7 (AuF2) and Sp1 (AuF3) in the 4+ charge state. The region selected for the major conformers are marked in red, while the region selected for the minor conformer is marked in blue. For NCp7 (AuF2), the overlap of the third conformer (marked with an arrow) with the major conformer precluded its isolation

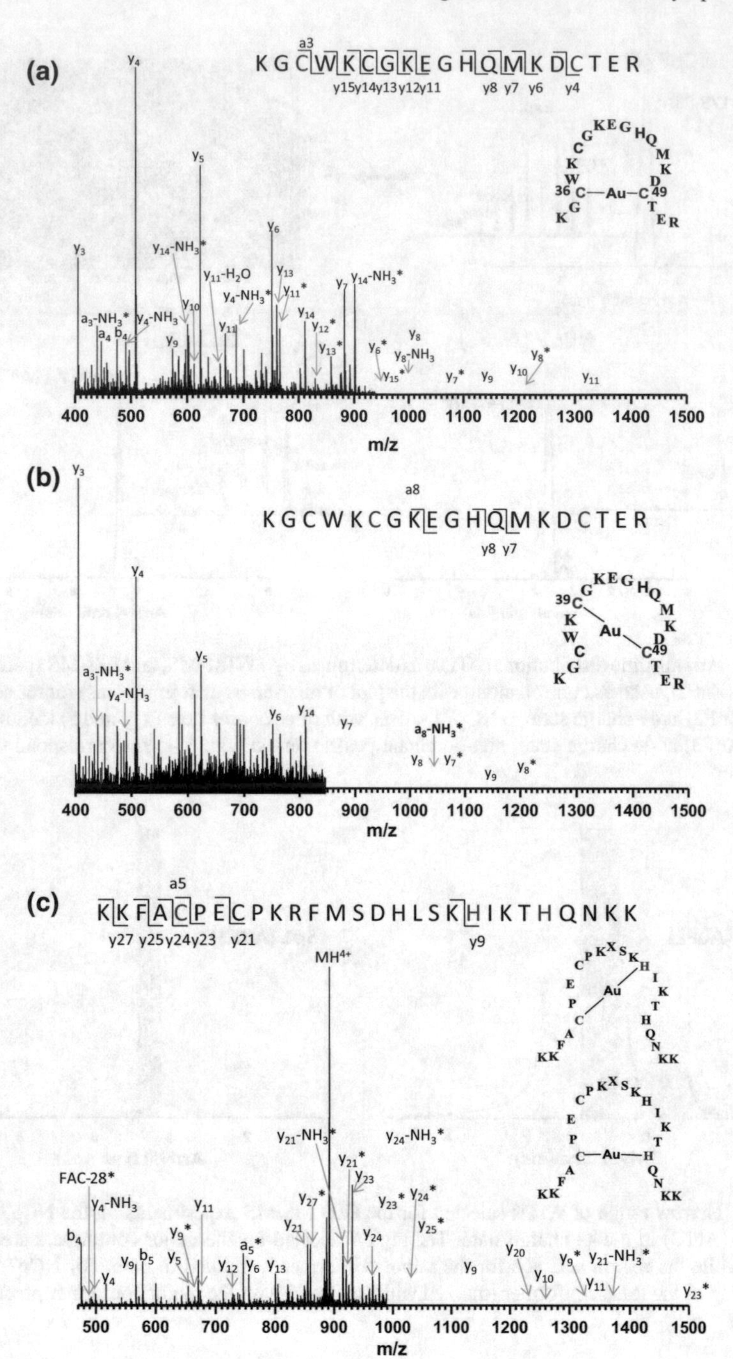

Fig. 2.4 Collision induced dissociation fragment spectra of the **a** major and **b** minor conformers in NCp7 AuF2 4+ and **c** the only conformer in Sp1 AuF3 4+ peaks (isolated regions are highlighted in Fig. 2.3). Proposed structures of each conformer based on the CID data are shown. a ion = b ion—carbonyl (C = O) [36]

Table 2.1 Identified gold-binding fragments arising from the MS/MS of the major conformer of NCp7 AuF2. "a_n" and "b_n" ions are N-terminal fragments while "y_n" ions are C-terminal fragments

Peptide fragment with Au(I)	Theoretical (m/z)*	Observed (m/z)*
a_3-NH_3 +Au(I)	440.07^+	440.10^+
y_{14}-NH_3 +Au(I)	600.88^{3+}	600.91^{3+}
y_4-NH_3 +Au(I)	687.15^+	687.19^+
y_{11} +Au(I)	765.26^{2+}	765.30^{2+}
y_{12} +Au(I)	829.31^{2+}	829.37^{2+}
y_{13} +Au(I)	857.82^{2+}	857.87^{2+}
y_{14}-NH_3 +Au(I)	900.81^{2+}	900.86^{2+}
y_6 +Au(I)	947.30^+	947.35^+
y_{15} +Au(I)	973.37^{2+}	973.42^{2+}
y_7 +Au(I)	1078.34^+	1078.39^+
y_8 +Au(I)	1206.40^+	1206.46^+

*most abundant isotopomer

Table 2.2 Identified unique gold-binding fragments arising from the MS/MS of the minor conformer (blue color) of NCp7 AuF2

Peptide fragment with Au(I)	Theoretical (m/z)*	Observed (m/z)*
a3-NH_3 +Au(I)	440.07^+	440.11^+
a8-NH_3 +Au	1042.37^+	1042.46^+
y7+Au(I)	1078.34^+	1078.42^+
y8+Au(I)	1206.40^+	1206.49^+

*most abundant isotopomer

is reasonable to propose that Au(I) coordinates in the CCHC zinc finger through Cys36—Cys49 crosslinking (Fig. 2.4a). The observation of aurated y_4-NH_3 in both the major and minor conformers of NCp7 AuF2 shows that Cys49 was metallated in both cases. The aurated a_3-NH_3 is mainly located in the dominant major conformer although some overlap with the minor conformer is observed.

The identified gold-binding fragments from the MS/MS spectrum of the minor conformer are shown in Table 2.2. The minor conformer is more compact ($\Omega = 163.57$ Å^2) and has good separation from the major conformer. Besides Cys49 auration, we can observe the aurated a_8-NH_3, y_7 and y_8 in the MS/M in total indicative of auration in either Cys36 or Cys39. Considering the above assignment of the major conformer, the minor conformer should have auration at Cys39 and Cys49 (Fig. 2.4b).

Cys_3His zinc fingers are especially susceptible to electrophilic attack and the C-terminal NCp7 ZnF2 is identified as being one of the most reactive, with Cys49 being the most labile site [21, 22]. The auration of Cys49 in both conformers confirms its greater reactivity. As stated, Cys49 is also the site of attack of NEM and treatment of

Table 2.3 Identified gold-binding fragments arising from the MS/MS of the only conformer of Sp1 C-terminal gold finger

Peptide fragment with Au(I)	Theoretical (m/z)*	Observed (m/z)*
FAC-28 + Au(I)	490.09^{+}	490.07^{+}
y_9 + Au(I)	665.31^{2+}	665.29^{2+}
a_5 + Au(I)	746.27^{+}	746.27^{+}
y_{27} + Au(I)	859.91^{4+}	859.88^{4+}
y_{21}-NH_3 + Au(I)	915.10^{3+}	915.06^{3+}
y_{21} + Au(I)	920.77^{3+}	920.75^{3+}
y_{23} + Au(I)	996.14^{3+}	996.10^{3+}
y_{24}-NH_3 + Au(I)	1024.81^{3+}	1024.75^{3+}
y_{24} + Au(I)	1030.48^{3+}	1030.44^{3+}
y_{25} + Au(I)	1054.49^{3+}	1054.44^{3+}
y_{21}-NH_3 + Au(I)	1372.15^{2+}	1372.10^{2+}
y_{23} + Au(I)	1493.70^{2+}	1493.63^{2+}

*most abundant isotopomer

infectious virus with NEM eliminated retroviral infectivity in a manner proportional to NEM concentration [23]. Replacement of Zn^{2+} in the full zinc finger NC eliminates the nucleic acid binding abilities of the peptide—crucial to its biological function in the HIV replication cycle [24]. These results strengthen our analogy between "organic" and Lewis acid electrophiles and suggest that, as well as useful probes of NC topography, the Au-based agents could provide a rich source of targeted anti-HIV agents based on a thiolate modification strategy [5, 25, 26].

For {AuF} derived from Sp1 (ZnF3), only one conformation was observed whose MS/MS spectrum is shown in Fig. 2.4c. The observation of aurated a_5 showed that Cys5 is a gold-binding site while the aurated y_9 is indicative of auration in either His 20 or His 24. The observation of aurated y_{21}, $y_{23–25}$, y_{27} is also consistent with the binding at His 20/His 24 and Cys5. The identified gold-binding fragments from the MS/MS spectrum are shown in Table 2.3. The linear Cys-Au-Cys was discarded based on MS/MS assignment.

Mass spectrometry and tandem MS have been used successfully to characterize gold-peptide/protein interactions but cannot assign specific binding sites in multi-receptor systems nor suggest the existence of conformers [27]. To our knowledge, this is the first demonstration that {AuF} can exist in structurally distinct conformations and represents a novel use of IM in protein structure analysis complementary to other spectroscopic and X-ray techniques [3, 4, 27–30]. Ion Mobility was used to study changes in the tertiary structure of ubiquitin upon metallation with cisplatin and revealed the presence of up to three different conformations of the Ub–{$Pt(NH_3)_2$} mono-adduct [31, 32]. Ruthenium(II)-peptide interactions studied by IM highlighted the important role played by the ligand in determining the shape of the adduct formed [33]. Furthermore, the conformational diversity of cancer-associated mutations in the zinc-bound p53 tumor suppressor protein has been probed by ion-mobility mass spectrometry [34]. A further relevant example is the assignment of Cu coordination sites on a methanobactin model peptide with potential Cys_2His_2 binding sites analogous to the Sp1 ZnF3 case [35].

2.4 Conclusions

Here we demonstrate the utility of ion mobility combined with MS/MS for the study of zinc finger protein/metal drug interactions. The ion mobility experiments are rapid (ms time-scale) and, when combined with MS/MS, will provide information not only on shape and conformer profile, but also on metalation selectivity and reactivity. TWIM allows separation of peptide species of the same mass-to-charge (m/z) ratio that exhibit multiple conformations. Therefore, when coupled to mass spectrometry TWIM-MS leads to very specific and complementary information about macromolecules, such as the crosslinking residues coordinated to Au different conformers of a protein (Scheme 2.1). In this specific case, the ability to identify precisely

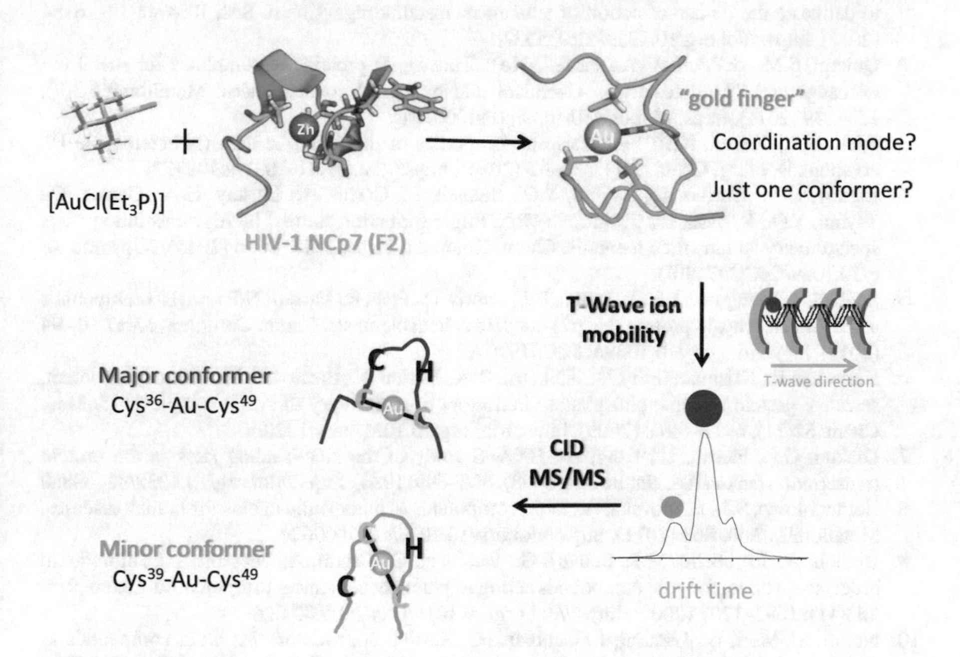

Scheme 2.1 Experimental route for separation of the different conformers of AuFs by TWIM-MS followed by CID

the nature of the auration and zinc displacement on the protein elucidates the details whereby DNA/RNA recognition is abrogated, considered essential in the inhibition of antiviral activity by these drugs. The screening of new metal-based anticancer or antiviral agents would benefit from quantitative information on the shape changes induced upon zinc finger protein metalation and the selectivity and reactivity of metalation.

Appendix

MS/MS Fragment Lists

See Tables 2.1, 2.2 and 2.3.

References

1. Hartinger, C.G., Groessl, M., Meier, S.M., Casini, A., Dyson, P.J., Tavernelli, I., Keppler, B.K., Jaehde, U., Messori, L., Messori, L., et al.: Application of mass spectrometric techniques to delineate the modes-of-action of anticancer metallodrugs. Chem. Soc. Rev. **42**(14), 6186 (2013). https://doi.org/10.1039/c3cs35532b
2. Quintal, S.M., dePaula, Q.A., Farrell, N.P.: Zinc finger proteins as templates for metal ion exchange and ligand reactivity. Chemical and biological consequences. Metallomics **3**(2), 121–139 (2011). https://doi.org/10.1039/c0mt00070a
3. Spell, S.R., Farrell, N.P.: Synthesis and properties of the first $[Au(dien)(N\text{-heterocycle})]^{3+}$ compounds. Inorg. Chem. **53**(1), 30–32 (2014). https://doi.org/10.1021/ic402772j
4. Laskay, Ü.A., Garino, C., Tsybin, Y.O., Salassa, L., Casini, A., Laskay, U.A., Garino, C., Tsybin, Y.O., Salassa, L., Casini, A.: Gold finger formation studied by high-resolution mass spectrometry and in silico methods. Chem. Commun. **51**(9), 1612–1615 (2015). https://doi.org/10.1039/C4CC07490D
5. Spell, S.R., Mangrum, J.B., Peterson, E.J., Fabris, D., Ptak, R., Farrell, N.P.: Au(III) compounds as HIV nucleocapsid protein (NCp7)–nucleic acid antagonists. Chem. Commun. **53**(1), 91–94 (2017). https://doi.org/10.1039/C6CC07970A
6. Karver, M.R., Krishnamurthy, D., Kulkarni, R.A., Bottini, N., Barrios, A.M.: Identifying potent, selective protein tyrosine phosphatase inhibitors from a library of Au(I) complexes. J. Med. Chem. **52**(21), 6912–6918 (2009). https://doi.org/10.1021/jm901220m
7. Diakun, G.P., Fairall, L., Klug, A.: EXAFS study of the zinc-binding sites in the protein transcription factor IIIA. Nature **324**(6098), 698–699 (1986). https://doi.org/10.1038/324698a0
8. Berners-Price, S.J., Filipovska, A.: Gold compounds as therapeutic agents for human diseases. Metallomics **3**(9), 863 (2011). https://doi.org/10.1039/c1mt00062d
9. Bindoli, A., Rigobello, M.P., Scutari, G., Gabbiani, C., Casini, A., Messori, L.: Thioredoxin reductase: a target for gold compounds acting as potential anticancer drugs. Coord. Chem. Rev. **253**(11), 1692–1707 (2009). https://doi.org/10.1016/j.ccr.2009.02.026
10. Nobili, S., Mini, E., Landini, I., Gabbiani, C., Casini, A., Messori, L.: Gold compounds as anticancer agents: chemistry, cellular pharmacology, and preclinical studies. Med. Res. Rev. **30**(3), 550–580 (2009). https://doi.org/10.1002/med.20168
11. Ott, I.: On the medicinal chemistry of gold complexes as anticancer drugs. Coord. Chem. Rev. **253**(11), 1670–1681 (2009). https://doi.org/10.1016/j.ccr.2009.02.019

12. Serebryanskaya, T.V., Lyakhov, A.S., Ivashkevich, L.S., Schur, J., Frias, C., Prokop, A., Ott, I., Rubbiani, R., Wahrig, B., Ott, I., et al.: Gold(I) thiotetrazolates as thioredoxin reductase inhibitors and antiproliferative agents. Dalton Trans. **44**(3), 1161–1169 (2015). https://doi.org/10.1039/C4DT03105A
13. Larabee, J.L., Hocker, J.R., Hanas, J.S.: Mechanisms of aurothiomalate-Cys2His2 zinc finger interactions. Chem. Res. Toxicol. **18**(12), 1943–1954 (2005). https://doi.org/10.1021/tx0501435
14. Handel, M.L., DeFazio, A., Watts, C.K., Day, R.O., Sutherland, R.L.: Inhibition of DNA binding and transcriptional activity of a nuclear receptor transcription factor by aurothiomalate and other metal ions. Mol. Pharmacol. **40**(5), 613–618 (1991)
15. Chertova, E.N., Kane, B.P., McGrath, C., Johnson, D.G., Sowder, R.C., Arthur, L.O., Henderson, L.E.: Probing the topography of HIV-1 nucleocapsid protein with the alkylating agent N-ethylmaleimide. Biochemistry **37**(51), 17890–17897 (1998). https://doi.org/10.1021/bi980907y
16. Sechi, S., Chait, B.T.: Modification of cysteine residues by alkylation. A tool in peptide mapping and protein identification. Anal. Chem. **70**(24), 5150–5158 (1998). https://doi.org/10.1021/ac9806005
17. Mendoza, V.L., Vachet, R.W.: Probing protein structure by amino acid-specific covalent labeling and mass spectrometry. Mass Spectrom. Rev. **28**(5), 785–815 (2009). https://doi.org/10.1002/mas.20203
18. Shaw, C.F.: Gold-based therapeutic agents. Chem. Rev. **99**(9), 2589–2600 (1999). https://doi.org/10.1021/cr980431o
19. Abbehausen, C., Peterson, E.J., De Paiva, R.E.F., Corbi, P.P., Formiga, A.L.B., Qu, Y., Farrell, N.P.: Gold(I)-phosphine-N-heterocycles: biological activity and specific (ligand) interactions on the C-terminal HIVNCp7 zinc finger. Inorg. Chem. **52**(19), 11280–11287 (2013). https://doi.org/10.1021/ic401535s
20. Hu, W., Luo, Q., Wu, K., Li, X., Wang, F., Chen, Y., Ma, X., Wang, J., Liu, J., Xiong, S., et al.: The anticancer drug cisplatin can cross-link the interdomain zinc site on human albumin. Chem. Commun. **47**(21), 6006 (2011). https://doi.org/10.1039/c1cc11627d
21. Maynard, A.T., Huang, M., Rice, W.G., Covell, D.G.: Reactivity of the HIV-1 nucleocapsid protein p7 zinc finger domains from the perspective of density-functional theory. Proc. Natl. Acad. Sci. U. S. A. **95**(20), 11578–11583 (1998). https://doi.org/10.1073/pnas.95.20.11578
22. Maynard, A.T., Covell, D.G.: Reactivity of zinc finger cores:? Analysis of protein packing and electrostatic screening. J. Am. Chem. Soc. **123**(6), 1047–1058 (2001). https://doi.org/10.1021/ja0011616
23. Morcock, D.R., Thomas, J.A., Gagliardi, T.D., Gorelick, R.J., Roser, J.D., Chertova, E.N., Bess, J.W., Ott, D.E., Sattentau, Q.J., Frank, I., et al.: Elimination of retroviral infectivity by N-ethylmaleimide with preservation of functional envelope glycoproteins. J. Virol. **79**(3), 1533–1542 (2005). https://doi.org/10.1128/JVI.79.3.1533-1542.2005
24. Levin, J.G., Guo, J., Rouzina, I., Musier-Forsyth, K., Rouzina, I., Musier-Forsyth, K.: Nucleic acid chaperone activity of HIV-1 nucleocapsid protein: critical role in reverse transcription and molecular mechanism. Prog. Nucleic Acid Res. Mol. Biol. **80**(05), 217–286 (2005). https://doi.org/10.1016/S0079-6603(05)80006-6
25. Mori, M., Kovalenko, L., Lyonnais, S., Antaki, D., Torbett, B.E., Botta, M., Mirambeau, G., Mély, Y.: Nucleocapsid protein: a desirable target for future therapies against HIV-1. Curr. Top. Microbiol. Immunol. **389**, 53–92 (2015). https://doi.org/10.1007/82_2015_433
26. Garg, D., Torbett, B.E.: Advances in targeting nucleocapsid–nucleic acid interactions in HIV-1 therapy. Virus Res. **193**, 135–143 (2014). https://doi.org/10.1016/j.virusres.2014.07.004
27. Lee, J., Jayathilaka, L.P., Gupta, S., Huang, J.-S., Lee, B.-S.: Gold ion-angiotensin peptide interaction by mass spectrometry. J. Am. Soc. Mass Spectrom. **23**(5), 942–951 (2012). https://doi.org/10.1007/s13361-011-0328-0
28. Zou, J., Taylor, P., Dornan, J., Robinson, S., Walkinshaw, M., Sadler, P.: First crystal structure of a medicinally relevant gold protein complex: unexpected binding of $[Au(PEt3)]^+$ to histidine. Angew. Chemie **39**(16), 2931–2934 (2000). https://doi.org/10.1002/1521-3773(20000818)39:16%3c2931:AID-ANIE2931%3e3.0.CO;2-W

29. Franzman, M.A., Barrios, A.M.: Spectroscopic evidence for the formation of goldfingers. Inorg. Chem. **47**(10), 3928–3930 (2008). https://doi.org/10.1021/ic800157t
30. Jacques, A., Lebrun, C., Casini, A., Kieffer, I., Proux, O., Latour, J.-M., Sénèque, O.: Reactivity of Cys 4 zinc finger domains with gold(III) complexes: insights into the formation of "gold fingers". Inorg. Chem. **54**(8), 4104–4113 (2015). https://doi.org/10.1021/acs.inorgchem.5b00360
31. Williams, J.P., Phillips, H.I.A., Campuzano, I., Sadler, P.J.: Shape changes induced by N-terminal platination of ubiquitin by cisplatin. J. Am. Soc. Mass Spectrom. **21**(7), 1097–1106 (2010). https://doi.org/10.1016/j.jasms.2010.02.012
32. Williams, J.P., Brown, J.M., Campuzano, I., Sadler, P.J., Sadler, P.J., Clausen, H., Johnsen, A.H., Zubarev, R.A., Dawson, A., Aird, R.E., et al.: Identifying drug metallation sites on peptides using electron transfer dissociation (ETD), collision induced dissociation (CID) and ion mobility-mass spectrometry (IM-MS). Chem. Commun. **46**(30), 5458 (2010). https://doi.org/10.1039/c0cc00358a
33. Murray, B.S., Menin, L., Scopelliti, R., Dyson, P.J.: Conformational control of anticancer activity: the application of arene-linked dinuclear ruthenium(II) organometallics. Chem. Sci. **5**(6), 2536 (2014). https://doi.org/10.1039/c4sc00116h
34. Jurneczko, E., Cruickshank, F., Porrini, M., Clarke, D.J., Campuzano, I.D.G., Morris, M., Nikolova, P.V., Barran, P.E.: Probing the conformational diversity of cancer-associated mutations in p53 with ion-mobility mass spectrometry. Angew. Chemie Int. Ed. **52**(16), 4370–4374 (2013). https://doi.org/10.1002/anie.201210015
35. Choi, D., Alshahrani, A.A., Vytla, Y., Deeconda, M., Serna, V.J., Saenz, R.F., Angel, L.A.: Redox activity and multiple copper(I) coordination of 2His-2Cys oligopeptide. J. Mass Spectrom. **50**(2), 316–325 (2015). https://doi.org/10.1002/jms.3530
36. Papayannopoulos, I.A.: The interpretation of collision-induced dissociation tandem mass spectra of peptides. Mass Spectrom. Rev. **14**(1), 49–73 (1995). https://doi.org/10.1002/mas.1280140104

Chapter 3
Probing Gold: X-Ray Absorption Spectroscopy

3.1 Introduction

In an effort to obtain further information on the systems addressed in Part I (interaction of Au(I) compounds with ZnF proteins), we discussed in Part I—Chap. 2 the approach of using TWIM-MS to observe and distinguish among different conformers of Au(I)-protein adducts based mainly on properties of the proteins. Here, our focus shifts towards probing directly the metal. X-ray Absorption Spectroscopy (XAS) is appropriate for structural studies due to its capability to specifically probe the geometry around the metal absorber. Additionally, XAS spectra are sensitive to the electronic density on the mental center. As consequence, it represents an extremely powerful technique that is able to distinguish between oxidation states and coordination spheres of a metal complex, providing insights on the electronic and geometric structural details of the protein bound metal centers.

In this chapter, the interaction of [AuCl(Et_3P)] (**I-1**) and auranofin (**I-3**) with zinc finger peptides was investigated using Au L_3-edge XAS measurements. To further support our conclusions, the data obtained for the reaction products were compared to the XAS spectra of selected model compounds and also to TD-DFT-calculated spectra. The viral NCp7 ZnF2 was compared to the human transcription factor Sp1 ZnF3. These represent two different families of ZnF proteins, NCp7 being the-Cys-X_2-Cys-X_4-His-X_4-Cys—(C_2HC) motif of the HIV-1, and the Cys-X_2-Cys-X_{12}-His-X_3-His (C_2H_2) Zn coordination sphere of Sp1, as schematically represented in Fig. 3.1. We examine the effects of the zinc coordination sphere in dictating reactivity and delineating the intimate mechanisms of metal ions substitution in a ZnF template. In the case of Au(I) compounds, we confirm the Lewis acid electrophilic attack of [AuCl(Et_3P)] and auranofin after interaction with NCp7, resulting in a common final product with Cys-Au-PEt_3 coordination sphere. In Part I—Chap. 1, we

---

Parts of this chapter has been reproduced with permission from ACS. https://pubs.acs.org/doi/10.1021/acs.inorgchem.7b02406.

R. E. Ferraz de Paiva, *Gold(I,III) Complexes Designed for Selective Targeting and Inhibition of Zinc Finger Proteins*, Springer Theses,
https://doi.org/10.1007/978-3-030-00853-6_3

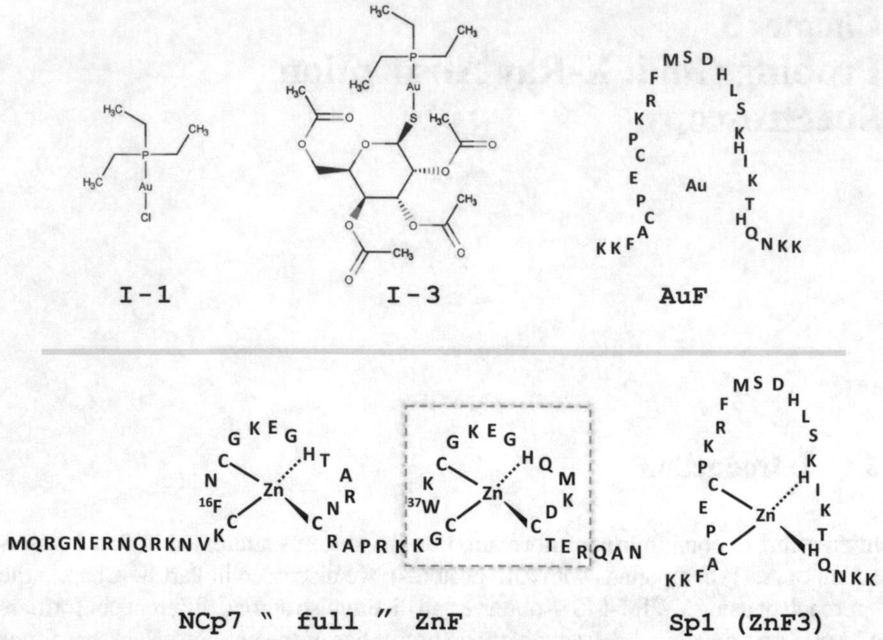

Fig. 3.1 Structural formulas of Au(I) compounds used in this study. The zinc finger model peptides NCp7 and Sp1 are also shown

demonstrated that the reaction product of [AuCl(Et_3P)] with Sp1 had a remarkably clean mass spectrum (Fig. 1.8). For that reason, we decided to purify and isolate the AuF obtained from that reaction and also study it using XAS. A Cys-Au-Cys coordination sphere is suggested for the the purified AuF studied here, as indicated by the electron rich gold center observed by XAS. As first discussed in Part I—Chap. 1, we also demonstrate here that the reactivity of [AuCl(Et_3P)] is also dependent on the ZnF core targeted.

3.2 Experimental

3.2.1 Synthesis and Zinc Finger Preparation

Compounds **I-1**, **I-3** and **M-6** were acquired from Sigma-Aldrich and used without further purification. The model compounds **M-1** to **M-5** (Fig. 3.5) were synthesized and purified according to literature procedures [1–4]. The crystal structures of all compounds, except **M-2** and **M-5**, have been already reported elsewhere [2, 3, 5–10]. Characterization of compounds **M-1** to **M-5** was performed by conventional spectroscopic techniques including ^{1}H, ^{13}C, ^{31}P NMR spectroscopy, mass spectrometry, elemental analysis, and infrared and UV spectroscopies, attesting the success of the synthesis and their integrity [1, 11].

The zinc finger models used in this study were the HIV-1 nucleocapsid protein ZnF2 and the human transcription factor Sp1 ZnF3. NCp7 ZnF2 and Sp1 ZnF3 were purchased from Aminotech Co. (São Paulo, Brazil), the full-length NCp7 ZnF was acquired from Invitrogen (USA). The apopeptides were checked by mass spectrometry and used as received. Both zinc fingers were prepared by dissolving sufficient mass of apopeptides in a 100 mmol L^{-1} solution of zinc acetate prepared in degassed water. The pH was adjusted to 7.2–7.4 using NH_4OH or HOAc if needed, leading to a solution with final concentration of 30 mmol L^{-1} of zinc finger. Sequences:

```
                     10        20
NCp7 ZnF2    KGCWKCGKEG HQMKNCTER
Sp1 ZnF3     KKFACPECPK RFMSDHLSKH IKTHQNKK
```

3.2.2 Sample Preparation

The solid samples of compounds **I-1** and **I-3** and model compound **M-1** to **M-6** were finely grounded and diluted in boron nitride to a maximum X-ray absorbance of about 1. They were then pressed into circular pellets of 13 mm diameter using a hydraulic press, placed in a plastic sample holder and covered with polyimide adhesive tape (Kapton) with about 40 μm thickness. Gold L_3-edge XAS measurements of the complexes **I-1** and **I-3** and model compound **M-1** to **M-6** were performed in solid state at XAFS1 beamline.

Stock solutions of $[AuCl(Et_3P)]$ (**I-1**) and auranofin (**I-3**) were prepared by dissolving the solid compounds in dmf to a final concentration of 100 mmol L^{-1}. For evaluating the interaction of the model compounds with the zinc fingers, a total sample volume of 10 μL of the Au(I) complexes **I-1** and **I-3** were prepared by mixing sufficient volumes of the solutions of model compounds to each zinc finger solution (NCp7(F2) and Sp1(F3)) in a molar ratio of 1:1.

The *gold finger* sample (AuF) from Sp1 (F3) was obtained incubating $[AuCl(Et_3P)]$ with Sp1-ZnF3 in a 1:1 molar ratio for 1 h. The reaction mixture was purified using a Waters 515 HPLC instrument with a reverse phase Phenomenex Jupiter C_{18} column (5 μm, 250 mm × 4.6 mm, 300 Å) and a 100 μL sample injection loop. Flow rate of 1 mL/min was used for all the experiments. The following gradient was used: 0–30 min, 10–80% B:A where A is H_2O with 0.1% TFA and B is acetonitrile with 0.1% TFA. UV detection wavelengths were 280 nm and 220 nm. The HPLC run is given in Fig. 3.6. The composition of the isolated fraction was further characterized by ESI-MS. The eluent from the collected sample was lyophilized and the solid was redissolved in ammonium phosphate buffer prior to XAS measurements. A sample with final concentration of 48 μmol L^{-1} was used.

In each XAS measurement of the samples in solution, about 3 μL of the prepared solutions of the zinc fingers and the Au(I) interaction products (**I-1** and **I-3**) were placed in a plastic sample holder, covered with the same 40 μm thick Kapton adhesive and frozen in a closed-cycle liquid helium cryostat. The solution samples were kept below 50 K during these measurements. Inspection of fast XANES scans revealed no signs of radiation damage, and the first and last scans of each data set used in the averages were identical. Gold L_3-edge X-ray absorption spectra of the samples in solution were acquired at the XAFS2 beamline [12].

Data averaging, background subtraction and normalization were done using standard procedures using the ATHENA package [13].

3.2.3 XAFS1 and XAFS2 Beamlines

XAFS1 an XAFS2 beamlines are located at the Brazilian Synchrotron Light Laboratory (CNPEM/LNLS) [12, 14]. At the XAFS1 beamline the incident energy was selected by a channel-cut monochromator equipped with a Si(111) crystal, and at the XAFS2 beamline a double-crystal, fixed-exit monochromator was used. The beam size at the sample was approximately 2.5×0.5 mm^2 (hor. x vert.) at XAFS1 and 0.4×0.4 mm^2 at XAFS2, with an estimated X-ray flux of 10^8 ph/s (XAFS1) and 10^9 ph/s (XAFS2). The incoming X-ray energy was calibrated by setting the maximum of the first derivative of L_3-edge of a gold metal foil to 11,919 eV. The XAS spectra of the solid samples (complexes **I-1** and **I-3** and model compound **M-1** to **M-6**) were collected in conventional transmission mode using ion chambers filled with a mixture of He and N_2. In the case of the samples in solution, fluorescence mode detection was used. The fluorescence signal was recorded using a 15-element Ge solid-state detector (model GL0055S—Canberra Inc.) by setting an integrating window of about 170 eV around the Au $L\alpha_1$ and $L\alpha_2$ emission lines (9713.3 eV and 9628.0 eV, respectively).

3.2.4 TD-DFT

Calculations were performed on all Au(I) compounds (**I-1** and **I-3** and model compound **M-1** to **M-6**) to see if our protocol could reproduce the trends in energy shifts and intensities. TD-DFT-obtained L_3-edges are shown in Appendix of this chapter, in an overall good agreement with the experimental data. Furthermore, theoretical models were created to model the Et_3P-Au(I)-Cys (**T-1**), Cys-Au(I)-Cys (**T-2**), His-Au(I)-His (**T-3**) and Cys-Au(I)-His (**T-4**) motifs. All calculations were performed using the ORCA quantum chemistry code, version 3.0.3 [15]. All molecules were optimized at the PBE0/def2-TZVP level of theory using the def2-ECP effective core potential [16]. TD-DFT calculations (using the Tamm-Dancoff approximation) [17] as implemented in ORCA were performed with the PBE0 [18–20] functional using an all-electron scalar relativistic Douglas-Kroll-Hess Hamiltonian [21–23] with the DKH-def2-TZVP-SARC basis set [24]. The Au 2p to valence excitations were performed by only allowing excitations from the Au 2p donor orbitals to all possible virtual orbitals of the molecule (only limited by the number of calculated roots). Intensities include electric dipole, magnetic dipole and quadrupole contributions. The RIJCOSX approximation [25, 26] was used to speed up the Coulomb and exchange integrals in both geometry optimizations and TD-DFT calculations.

3.3 Results and Discussion

Eight Au(I) complexes representing different X-Au-Y coordination spheres (with the donor atoms X, Y = S, P, Cl, N and/or C) were selected (Fig. 3.5) to serve as models for different relevant coordination spheres of Au(I), functioning as analogues to the interaction products between the [AuCl(Et_3P)] and auranofin with the target zinc finger proteins.

In general, the XAS spectra of the Au(I) model compounds present the expected white line peak in the 11,923–11,925 eV range. L3-edge whitelines have been seen for Au(I) [$d^{10}6\,s^0$ valence configuration], suggesting some unoccupied d-orbital character to be present or perhaps being due to other transitions (see TD-DFT discussion). In a first approximation, the intensity of the L_3-edge XAS white line is proportional to the d-electron count allowing the inference of the metal oxidation state [27]. A summary of the spectroscopic features observed for each model compound is given in Part I—Appendix in Chap. 3 (Fig. 3.5 and Table 3.1). Figure 3.3 shows the TD-DFT spectra obtained for the model Au(I) compounds, in comparison to the experimental data.

3.3.1 Interaction with NCp7 ZnF2

Figure 3.2 shows the Au L_3-edge XANES spectra of the Au(I) compounds **I-1**, **I-3** and **M-5** and the interactions of **I-1** with the NCp7 and Sp1 zinc fingers. Compound **I-1** presents a Au XANES spectrum similar to the other Au(I) compounds with P-Au-Y coordination (where Y can be Cl, N or S), with a white line peak located at 11,924.2 eV and about 0.83 normalized units. After interaction with NCp7, the XANES of the resulting product **I-1**+NCp7 changes slightly: the broad and weak white line sharpens and gains intensity, peaking at 11,925.7 eV and 0.97 normalized units. This final spectrum is almost identical to that of auranofin (**I-3**), which has a peak at the same energy and only slightly more intense (0.98 units), indicating similar geometry and electronic configuration between these species. In Fig. 3.2 the XANES spectrum of free auranofin (compound **I-3**) is remarkably similar to that of the reaction **I-1**+NCp7 and **I-3**+NCp7. As the near-edge XAS region is sensitive to both the atomic arrangement and the electronic structure around the metal [28–31], this indicates that the products formed present the same coordination sphere as found in auranofin, i.e., a

Table 3.1 Edge position, oxidation state and approximated site symmetry from studied compounds and Au(0) as reference

Compound		Edge position (eV)	Oxid. St.	Site symmetry (around Au atom)
#	Name			
Au(0)	Au foil	11919.0	0	–
I-1	$[AuCl(Et_3P)]$	11920.0	1+	$C_{\infty v}$
I-3	auranofin	11920.4	1+	$C_{\infty v}$
M-1	$[AuCl(Ph_3P)]$	11920.2	1+	$C_{\infty v}$
M-2	$[Au(dmap)(Ph_3P)]^+$	11920.7	1+	$C_{\infty v}$
M-3	$[Au(mtz)(Ph_3P)]$	11920.4	1+	$C_{\infty v}$
M-4	$[Au(CN)(Ph_3P)]$	11922.3	1+	$C_{\infty v}$
M-5	[Au(*N*-Ac-Cys)]	11919.7	1+	$D_{\infty h}$
M-6	$[Au(CN)_2]^-$	11921.5	1+	$D_{\infty h}$
Reaction	Name	Edge position (eV)	Oxid. St.	Coordination sphere
I-1+NCp7	$[AuCl(Et_3P)]$ +NCp7 ZnF2	11920.8	1+	P-Au-S(Cys)
I-3+NCp7	auranofin + NCp7 ZnF2	11920.9	1+	P-Au-S(Cys)
I-1+Sp1	AuF prepared from Sp1 ZnF3	11920.3	1+	(Cys)S-Au-S(Cys)

P-Au-S coordination. The same also holds true regarding the electronic configuration of these compounds, being all formally Au(I). The similarity in coordination sphere between auranofin and its product preclude conclusions on substitution at this early time point, but the MS spectrum obtained for **I-3**+NCp7 shows the presence of the $\{Et_3PAu\}$ moiety bound to the peptide, suggesting a possible replacement of the thiosugar by apo-NCp7 (Fig. 1.26). The proposed coordination environment for the reaction products of the Au(I) models with NCp7 is further supported by the inspection of the EXAFS (Extended X-ray Absorption Fine Structure) region of the spectra. The EXAFS of **I-1**+NCp7 and **I-3**+NCp7 are virtually superimposable with that of free auranofin up to $k = 10\ Å^{-1}$ (Fig. 3.8). As the crystal structure of the latter is known [10], it is possible to conclude that the gold atoms in these reaction products have identical neighborhoods, i.e., P-Au-S. This confirms our previous findings that the reaction of **I-1** with a NCp7 zinc finger results in a long-lived Et_3PAu finger species (see Part I—Chap. 1), while also revealing that the reaction of auranofin with NCp7 also produces the same species [32]. Travelling Wave Ion Mobility spectra of the {AuF} species formed from **I-1**+NCp7 indicated the presence of Cys-Au-Cys conformers (Part I—Chap. 2), a consequence of the multiple possible cysteine binding sites [33]. The formation of these conformers can now be readily understood by first the formation of $\{Au(Et_3P)\text{-}F\}$ species as confirmed here followed by loss of the Et_3P ligand and formation of a second Au-Cys bond.

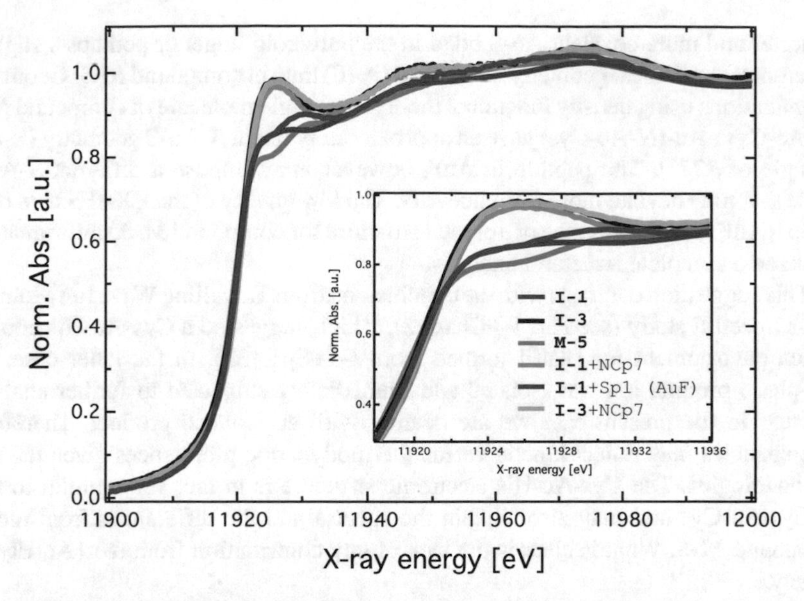

Fig. 3.2 XANES spectra of the compounds **I-1** and **I-3** and of the products of the their interaction with NCp7 and Sp1. Compound **M-5** (S-Au-S coordination) is also shown for comparison. The inset shows a zoom-in of the white line region, highlighting the spectral changes after interaction with the ZnFs

3.3.2 The Purified AuF and Interactions with Sp1

The XANES of the purified AuF obtained by the reaction of [AuCl(Et_3P)] with Sp1 ZnF3, which was isolated and purified by HPLC (from here on referred to as **I-1**+Sp1(AuF)) differs considerably from that of compound **I-1**, as shown in Figs. 3.2 and 3.4. The white line region of **I-1** contains a broad and weak peak, which splits in two in the interaction product **I-1**+Sp1(AuF), with an energy separation of about 5.5 eV. Moreover, the spectrum of **I-1**+Sp1(AuF) is clearly distinct from that of pure auranofin (**6**) and the interaction products **I-1**+NCp7 and **I-3**+NCp7, indicating that different species are formed for **I-1**+Sp1 and pointing to possible selectivity. The weak double feature in the white line of **I-1**+Sp1(AuF) (around 11,924–11,929 eV) is also evident in the XANES of compound **M-5** ([Au(*N*-Ac-Cys)]), which contains an S-Au-S environment. In the post-edge region of **I-1**+Sp1(AuF) a single broad feature is present, contrasting with the two marked structures observed in the spectra of **I-3**, **I-3**+NCp7 and **I-1**+NCp7. Since a similar single feature is also observed in the spectrum of **M-5**, we can infer that the purified gold-finger has a similar structure as compound **M-5**, i.e., a S-Au-S coordination and formally Au(I) oxidation state. The EXAFS of **I-1**+Sp1(AuF) essentially overlaps with that of compound **M-5** up to $k = 8$ Å^{-1} (Fig. 3.9). The apparently slightly shorter bond distance in the first coordination shell of AuF suggested by EXAFS and the differences in the white line of **I-1**+Sp1(AuF) and compound **M-5**, could be explained by the formation of a shorter and more covalent Au-S bond in the pure gold finger or perhaps a slightly different S-Au-S local geometry in **I-1**+Sp1(AuF) than in compound **M-5**. Geometry optimizations using density functional theory of a single molecule of compound **M-5** (*N*-Ac-Cys)-Au-(*N*-Ac-Cys) gave an approximately linear S-Au-S geometry (S-Au-S angle of 177°). The peptide in AuF, however, may impose a different S-Au-S angle that may deviate more from linearity. The low quality of the EXAFS data of **I-1**+Sp1(AuF) and the absence of a crystal structure for compound **M-5**, unfortunately, hindered a complete structural analysis.

This suggestion contrasts with the conclusions from Travelling Wave Ion Mobility Mass Spectral study (see Part I—Chap. 2), which suggested a Cys-Au-His coordination environment for [AuF] formed from **I-1**+Sp1 [33]. In the latter case, the gas-phase product ion was isolated and immediately subjected to further analysis whereas in the present case we are dealing with an isolated product. Therefore, discrepancies may reflect kinetic versus thermodynamic preferences given the two methodologies. The Cys-Au-His electronic structure is in fact very similar to that of Cys-Au-Cys and may also explain the spectra and the differences from model compound **M-5**. What is clear is the lack of any contribution from the {Au(Et_3P)} moiety.

3.3.3 TD-DFT

To provide a quantitative description of the interaction products of the Au(I) compounds with the ZnFs, we have employed DFT structural modeling and TD-DFT calculations of the Au L_3-edge XAS. TD-DFT calculations of metal K-edge XAS spectra have become popular and have been found to reproduce important pre-edge spectral features of metal complexes and cofactors well [34–40]. While L-edge spectra of first-row transition metals require more sophisticated wavefunction theory calculations that account for the difficult multiplet effects and spin-orbit coupling involving p- and d-shells, it has been found that a TD-DFT approach, neglecting both spin-orbit coupling and multiplet effects works well for L-edges of second-row transition metals such as molybdenum and ruthenium, where the multiplet effects should be small [31, 41]. For Au, a third-row transition metal, this approach would also be expected to work well.

The Au(I) compounds evaluated in this work had their structures optimized using DFT. More specifically, compound **M-5** was modeled as an almost linear Cys-Au-Cys model, while **I-1**, **I-3** and **M-3** had their structures optimized based on the available crystalline structures [2, 3, 10]. Optimized structures are shown in Fig. 3.7. Theoretical models of the interaction products were also created based on Et_3P-Au(I)-Cys (**T-1**), Cys-Au(I)-Cys (**T-2**), His-Au(I)-His (**T-3**) and Cys-Au(I)-His (**T-4**) motifs with Cys being modeled as N-acetyl-N-methylamide-cysteine and His residues modeled as 5-methylimidazole (Fig. 3.3). Optimizations and TD-DFT calculations were performed using a protocol described previously [41]. Additional details are given in the Experimental Section.

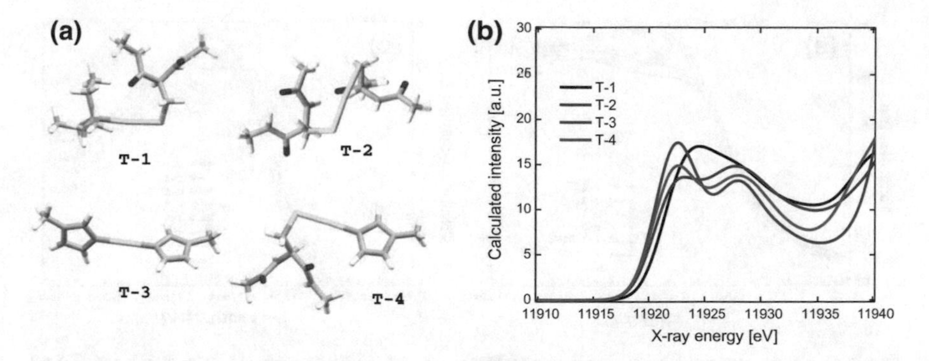

Fig. 3.3 **a** DFT optimized structures of the theoretical models used for the Au(I)-protein adducts. Cys was modeled as *N*-acetyl-*N*-methylamide-cysteine and His residues were modeled as 5-methylimidazole. **b** TD-DFT-calculated spectra of the theoretical Au(I) model compounds

Figure 3.4 shows a zoom-in of the white line region of the Au L_3-edge XANES of the Au(I) model compounds (**I-1**, **I-3**, **M-3** and **M-5**) and the selected interaction products **I-1**+NCp7 and **I-1**+Sp1(AuF), and their corresponding TD-DFT-calculated spectra, respectively. An energy shift of −465 eV was applied to the calculated spectra based on the correlation between experimental and calculated energies. The constant shift is necessary due to the sensitivity of the 2p core-orbital energies to relativistic effects as well as the approximate exchange-correlation potential, which are not perfectly described in our calculations. A broadening of 4.4 eV FWHM, based on the Au $2p_{3/2}$ core-hole lifetime, was used in all calculated spectra. We note that the TD-DFT-calculated spectra can only be expected to model the edge-region well, not the post-edge region of the XAS spectra. Overall, the calculated spectra nicely reproduce the observed experimental trends, both in terms of the energies and the intensity distribution (Fig. 3.4). Furthermore, TD-DFT can reveal the nature of the transitions involved. As these are all Au(I) model compounds with formally a full d-electron shell, from an electronic structure point of view, Au 2p → Au 6 s orbital excitations should be the first main contribution to the edge. Analysis of the transitions reveals that the Au 6 s orbital mixes strongly with the bound-ligand orbitals in these compounds and the excitations are better described as Au 2p → Au-L σ*. These excitations make up the main peak in the near edge region of the spectrum of all Au(I) compounds, but higher energy charge-transfer excitations into various empty ligand orbitals, in particular the phosphorus and sulfur d-shells, contribute as well and broaden the peak or result in a second peak, as in the case of compounds **I-1**, **M-5**, and Cys-Au-Cys (**T-2**). In the sulfur-containing Au-compounds (**M-3**, **M-5**, Et_3P-Au(I)-Cys (**T-1**) and Cys-Au(I)-Cys (**T-2**)), the Au 2p → Au-L σ* excitations are shifted slightly to higher energies, while the higher intensity of the peak compared to compound **I-1** can be attributed to additional Au 2p → S d excitations (with some mixing with Au 5d orbitals due to covalency), that arise due to the more covalent Au-S bond compared to the Au-Cl bond.

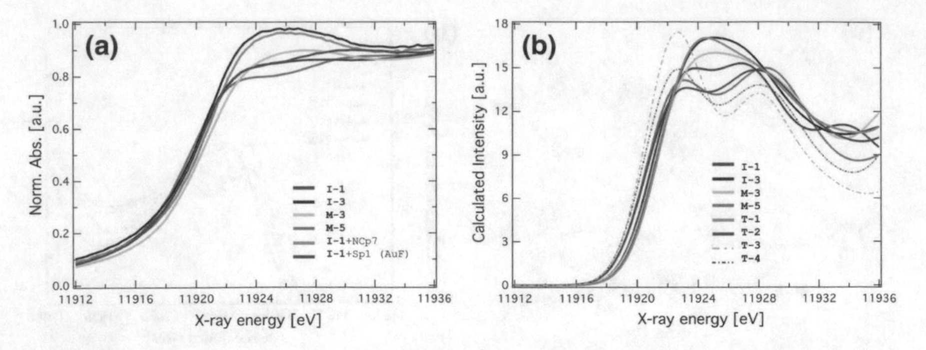

Fig. 3.4 **a** Experimental Au L_3-edge XANES spectra of Au(I) compounds **I-1**, **I-3**, **M-3** and **M-5**, and interaction products **I-1**+NCp7 and **I-1**+Sp1 (AuF). **b** TD-DFT-calculated L-edge XANES spectra of the same compounds and the proposed models Et_3P-Au-Cys (**T-1**) for **I-1**+NCp7 and Cys-Au-Cys (**T-2**) for **I-1**+Sp1. The TD-DFT-calculated spectra have been shifted in energy by −465 eV and broadened by 4.4 eV (FWHM)

Furthermore, the calculated spectrum of the simple theoretical models Et_3P-Au(I)-Cys (I-A, Fig. 3.3) is similar to that of compound **I-3**, in agreement with the experimental spectrum, again supporting our hypothesis that the **I-1**+NCp7 and **I-3**+NCp7 products consist of a P-Au-S local geometry, i.e., the Et_3P-Au motif binding to a single Cys residue within the peptide. In the case of interaction with Sp1, the comparison of the simulated spectra with the experimental ones indicates that Cys-Au(I)-Cys (**T-2**) is clearly a much better model for **I-1**+Sp1 than Et_3P-Au(I)-Cys (**T-1**) although the spectrum of the Cys-Au-His (**T-4**) is quite similar to that of **T-2**, with slightly different relative intensities. Theoretical model **T-2** and compound **M-5** are structurally very similar and this can be verified by comparing their calculated spectra. We also note that the lower energy of the phosphorus d-shell compared to the sulfur d-shell, results in a single broad high-intensity peak for the P-Au(I)-S compounds [**I-3** and Et_3P-Au(I)-Cys (theoretical model **T-1**)], while the S-Au-S compounds and S-Au-N compounds (**M-5**, **T-2** and **T-4**) give two peaks (See Fig. 6.3, Part 2—Chap. 6).

The overall picture in the case of Au(I) model compounds interacting with NCp7 and Sp1 ZnFs is that the gold atom maintains its oxidation state upon interaction. However, it presents a slightly higher electronic density on the metal center in the case of the interaction product with Sp1 when compared to NCp7. The stronger donor character of thiols might be responsible for the reduction of the metal center. The formation of different final products after interaction with Au(I) compounds, depending on the ZnF is evidenced and suggests different zinc displacement mechanisms for compound **I-1** when targeting NCp7 or Sp1. This is an important information for the design of more specific inhibitors. The observation of different products after interaction of **I-1** with the two zinc fingers is intriguing—why do we observe {AuF} as a dominant species? The result may reflect the greater reactivity of the Sp1 core with respect to Zn^{2+} affinity. The PR_3 adduct must be formed but further reaction on Sp1 may be too fast to be observed. In this respect it is of interest to note that binding constants for Zn^{2+} to the two peptides differ significantly with the NCp7 being held much more tightly [42, 43]. This interpretation is bolstered by observation of {(PPh_3)Au-F} adducts when the electrophile is the relatively more substitution-inert $[Au(PPh_3)(dmap)]^+$ [11]. Likewise the ESI-MS spectrum of **I-3** where the S-bound leaving group is more inert than Cl^- does show the presence of {$(Et_3$PAuF} species (see Fig. 1.26, Part I—Chap. 1).

To our knowledge, this is the first Au(I)+ZnF system evaluated by a combination of XAS and TD-DFT. Here, the element-specific technique of XAS combined with the TD-DFT calculations corroborate the hypothesis of electrophilic attack of the metal center (Au(I)) on the cysteine, with replacement of the chloride (compound **I-1**) or thiosugar (compound **I-3**) ligands, and the maintenance of the Et_3P group upon initial reaction with NCp7. Interestingly, this coordination environment has also been observed for the systems auranofin + bovine serum albumin and auranofin + apo-transferrin [44]. In contrast, the human transcription factor Sp1 presented a different reaction product when interacting with $[AuCl(Et_3P)]$, showing no evidence for the phosphine-bound intermediate, and generating species with S-Au-S or a S-Au-N-coordination.

3.4 Conclusions

A systematic study using a combination of XAS and TD-DFT calculations based on a selection of experimental and theoretical models allowed the assessment of the coordination environments of the gold complexes formed by the interaction of Au(I) compounds with different zinc fingers. Gold L_3-edge XAS allowed the identification of the interaction products of $[AuCl(Et_3P)]$ (**I-1**) and auranofin (**I-3**) with NCp7 ZnF2, which have S-Au-P coordination spheres, as detected for reaction products immediately frozen and analyzed. This observation confirms the mechanism based on the electrophilic attack of the Au(I) centers on the zinc-bound Cys residues, with maintenance of the coordinated phosphines in the resulting products. TD-DFT calculations provided more information based the comparison of theoretical models for Au(I) linear coordination (namely Et_3P-Au-Cys, Cys-Au-Cys, His-Au-His and Cys-Au-His,) with the experimental XAS data, supporting the proposed Et_3P-Au-S(Cys) coordination sphere for **I-1**+NCp and **I-3**+NCp7.

Furthermore, our results show that the **I-1**+Sp1(AuF) is different from the product formed when interacting **I-1** with NCp7. TD-DFT supports the conclusion of an (Cys)S-Au-S(Cys) coordination sphere for **I-1**+Sp1 (AuF), in agreement with a slightly more electron-rich gold center (Scheme 3.1).

In summary, we demonstrated that a combination of XAS and TD-DFT calculations can be a powerful tool allowing the elucidation of the coordination sphere and oxidation state of Au compounds upon interaction with two different ZnFs. We also demonstrated that the reactivity depends on properties of both the Au complex and the ZnF studied. The XAS results obtained and discussed here bring additional structural information providing means to distinguish between limiting mechanisms involving structurally distinct zinc fingers. Understanding the factors governing the

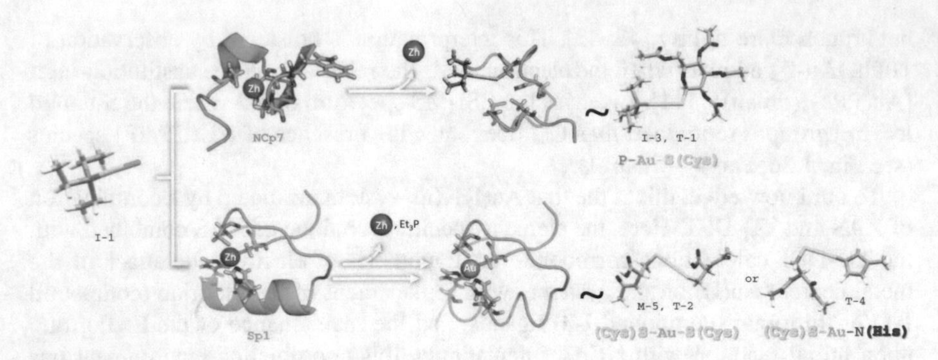

Scheme 3.1 Proposed mechanism for Zn(II) displacement caused by [AuCl(Et_3P)] (compound **I-1**) upon interaction with NCp7 and Sp1. The final coordination sphere of Au is highlighted in each case

metal replacement reaction will allow the rational design of new compounds with better inhibition properties towards ZnFs. The robust approach presented here, combining experimental XAS data with TD-DFT calculation can be extended to evaluate the interaction of a wide range of metallodrugs with target biomolecules.

Appendix

XAS of Au(I) Model Compound and TD-DFT Calculations

See Figs. 3.5, 3.6 and 3.7; Table 3.1.

EXAFS

See Figs. 3.8 and 3.9.

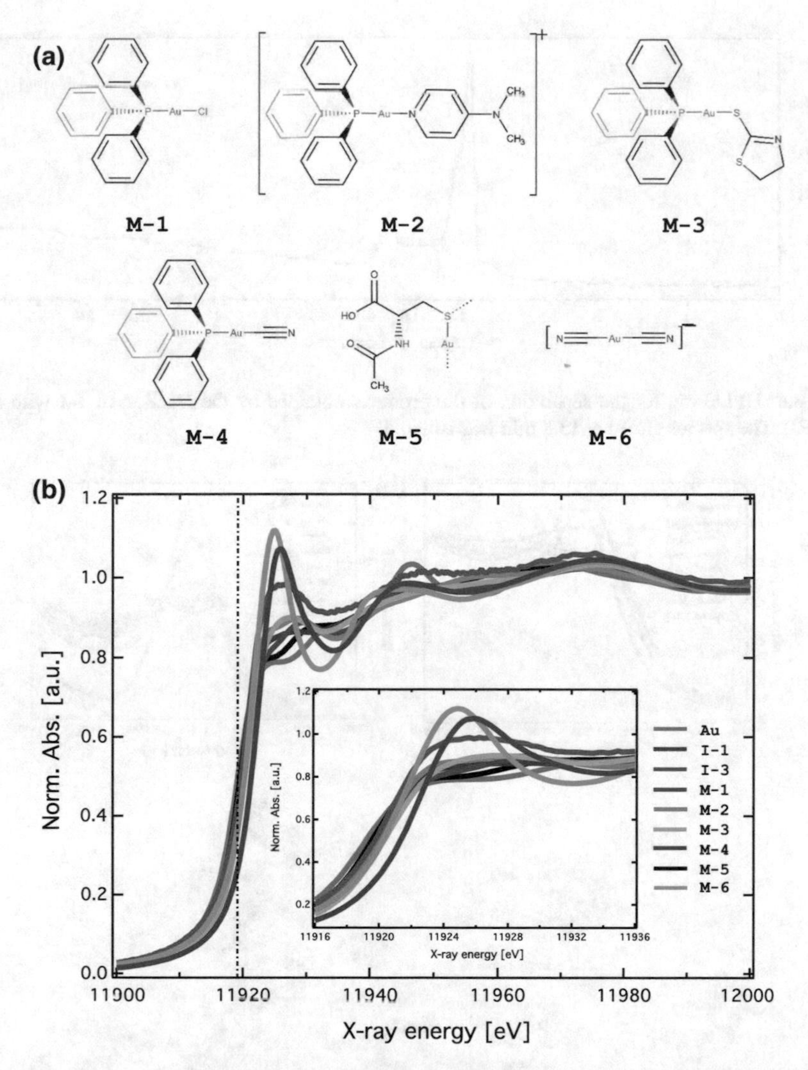

Fig. 3.5 **a** List of Au(I) model compounds used in this study. **b** Gold L_3 XANES spectra of the model complexes. The spectrum of Au foil (represented as Au) is also shown for comparison, and its edge position (19,919 eV) is marked as a dashed vertical line. Note the strong white line peak present in the Au(I) compounds **M-4** and **M-6** with CN^- ligands

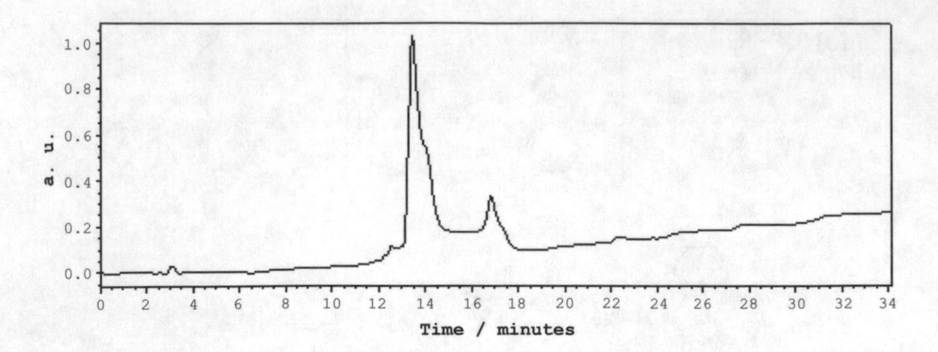

Fig. 3.6 HPLC run for the separation of the products obtained by the reaction of **I-1** with Sp1 (ZnF3). The species eluted at 13.8 min was isolated

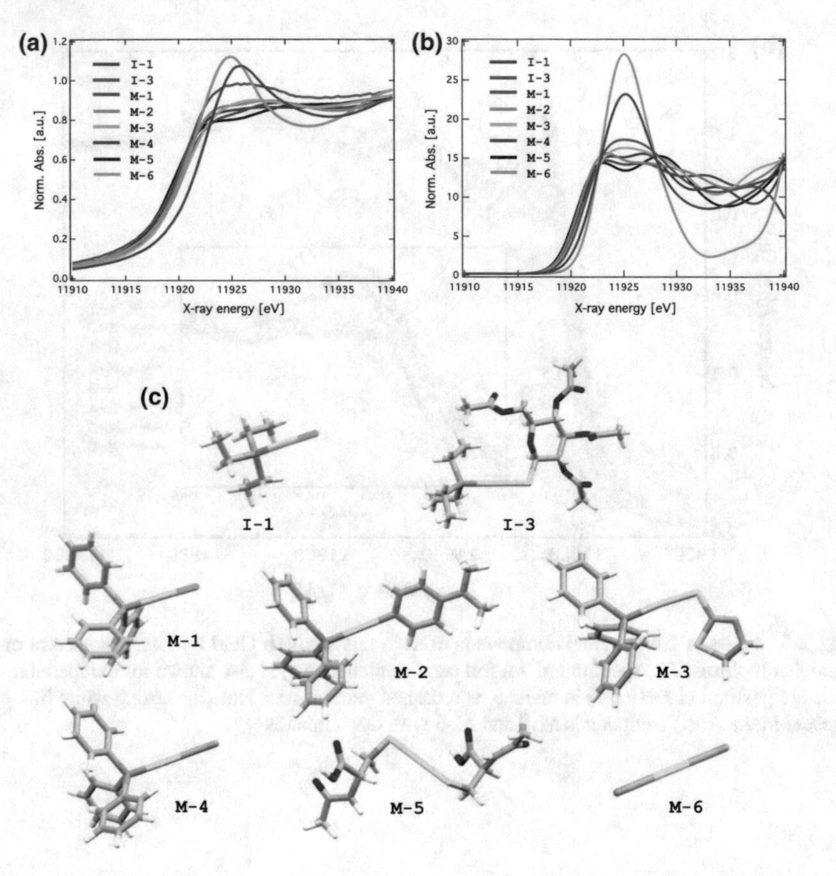

Fig. 3.7 **a** Experimental Au L_3-edge spectra Au(I) compounds **I-1, I-3** and models **M-1** to **M-6** compared to **b** TD-DFT-calculated L-edge spectra. The TD-DFT-calculated spectra have been shifted by 465 eV to lower energies. **c** DFT-optimized structures of the experimental Au(I) model compounds

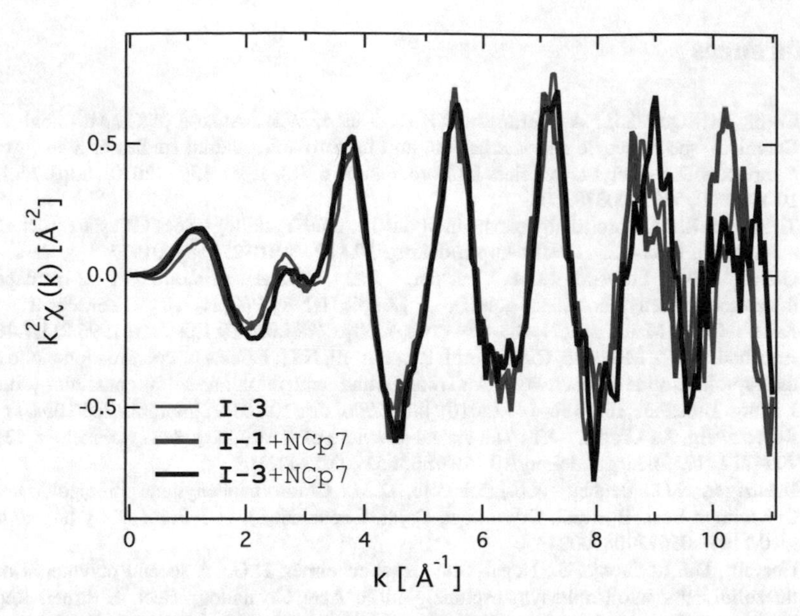

Fig. 3.8 Comparison of the k^2-weighted EXAFS of auranofin (**I-3**), the product **I-1**+NCp7 and the product **I-3**+NCp7

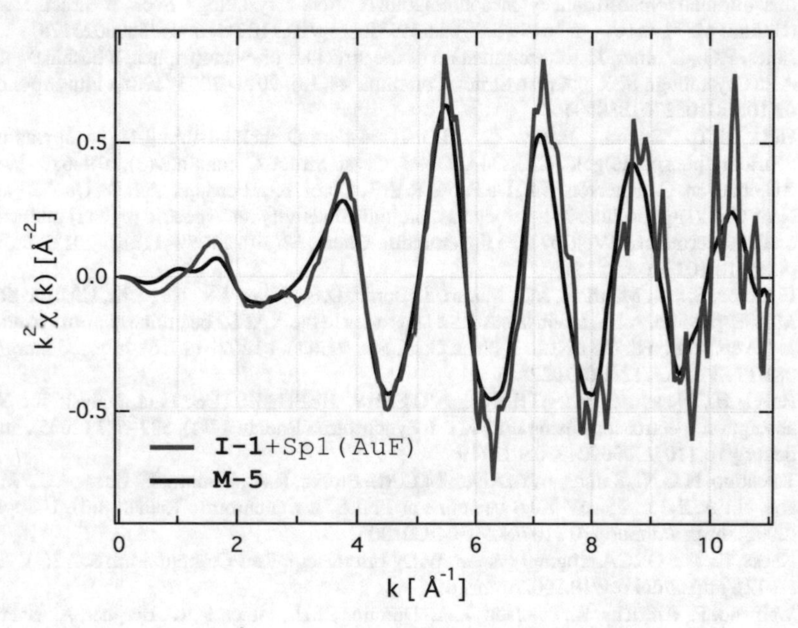

Fig. 3.9 Comparison of the k^2-weighted EXAFS of model compound **M-5**, with coordination sphere S-Au-S, and the isolated pure gold finger (Sp1)

References

1. Corbi, P.P., Quintão, F.A., Ferraresi, D.K.D., Lustri, W.R., Amaral, A.C., Massabni, A.C.: Chemical, spectroscopic characterization, and in vitro antibacterial studies of a new gold(I) complex with N-acetyl-L-cysteine. J. Coord. Chem. **63**(8), 1390–1397 (2010). https://doi.org/10.1080/00958971003782608
2. Tiekink, E.R.T. Chloro(triethylphosphine)gold(I). Acta Crystallogr. Sect. C Cryst. Struct. Commun. **45**(8), 1233–1234 (1989). https://doi.org/10.1107/s0108270189001915
3. Grant, T.A., Forward, J.M., Fackler, J.P.: Crystal structure of 2-mercapto-2-thiazoline(triphenylphosphine)-gold(I), $[Au(C_3H_4NS_2)P(C_6H_5)_3]$. Zeitschrift für Krist.—Cryst. Mater. **211**(7), 483–484 (1996). https://doi.org/10.1524/zkri.1996.211.7.483
4. Abbehausen, C., Manzano, C.M., Corbi, P.P., Farrell, N.P.: Effects of coordination mode of 2-mercaptothiazoline on reactivity of Au(I) compounds with thiols and sulfur-containing proteins. J. Inorg. Biochem. **165**, 136–145 (2016). https://doi.org/10.1016/j.jinorgbio.2016.05.011
5. Rosenzweig, A., Cromer, D.T.: The crystal structure of $KAu(CN)_2$. Acta Crystallogr. **12**(10), 709–712 (1959). https://doi.org/10.1107/S0365110X59002109
6. Baenziger, N.C., Bennett, W.E., Soborofe, D.M.: Chloro(triphenylphosphine)gold(I). Acta Crystallogr. Sect. B Struct. Crystallogr. Cryst. Chem. **32**(3), 962–963 (1976). https://doi.org/10.1107/s0567740876004330
7. Horvath, U.E.I., Cronje, S., Nogai, S.D., Raubenheimer, H.G.: A second polymorph of (2-thiazolidinethionato)(triphenylphosphine)gold(I). Acta Crystallogr. Sect. E Struct. Reports (Online) **62**(7), m1641–m1643 (2006). https://doi.org/10.1107/S1600536806023166
8. Horvath, U.E.I., Raubenheimer, H.G.: A third polymorph of (2-thiazolidinethionato)(triphenylphosphine)gold(I). Acta Crystallogr. Sect. E Struct. Reports (Online) **62**(7), m1644–m1645 (2006). https://doi.org/10.1107/S1600536806023178
9. Jones, P.G., Lautner, J.: Redetermination of the structure of cyano(triphenylphoshine)gold(I). Acta Crystallogr. Sect. C Cryst. Struct. Commun. **44**(12), 2091–2093 (1988). https://doi.org/10.1107/s0108270188008443
10. Hill, D.T., Sutton, B.M.: (2, 3, 4, 6-tetra-O-acetyl-1-thio-β-D-glucopyranosato-S)(triethylphosphine)gold, C20H34AuO9PS. Cryst. Struct. Commun. **9**(3), 679–686 (1980)
11. Abbehausen, C., Peterson, E.J., De Paiva, R.E.F., Corbi, P.P., Formiga, A.L.B., Qu, Y., Farrell, N.P.: Gold(I)-phosphine-N-heterocycles: biological activity and specific (ligand) interactions on the C-terminal HIVNCp7 zinc finger. Inorg. Chem. **52**(19), 11280–11287 (2013). https://doi.org/10.1021/ic401535s
12. Figueroa, S.J.A., Mauricio, J.C., Murari, J., Beniz, D.B., Piton, J.R., Slepicka, H.H., de Sousa, M.F., Espíndola, A.M., Levinsky, A.P.S.: Upgrades to the XAFS2 beamline control system and to the endstation at the LNLS. J. Phys: Conf. Ser. **712**(1), 012022 (2016). https://doi.org/10.1088/1742-6596/712/1/012022
13. Ravel, B., Newville, M.: ATHENA, ARTEMIS, HEPHAESTUS: data analysis for X-ray absorption spectroscopy using IFEFFIT. J. Synchrotron Radiat. **12**(4), 537–541 (2005). https://doi.org/10.1107/S0909049505012719
14. Tolentino, H.C.N., Ramos, A.Y., Alves, M.C.M., Barrea, R.A., Tamura, E., Cezar, J.C., Watanabe, N.: A, 2.3 to 25 keV XAS beamline at LNLS. J. Synchrotron Radiat. **8**(3), 1040–1046 (2001). https://doi.org/10.1107/S0909049501005143
15. Neese, F.: The ORCA program system. Wiley Interdiscip. Rev. Comput. Mol. Sci. **2**(1), 73–78 (2012). https://doi.org/10.1002/wcms.81
16. Weigend, F., Ahlrichs, R., Peterson, K.A., Dunning, T.H., Pitzer, R.M., Bergner, A.: Balanced basis sets of split valence, triple zeta valence and quadruple zeta valence quality for H to Rn: design and assessment of accuracy. Phys. Chem. Chem. Phys. **7**(18), 3297 (2005). https://doi.org/10.1039/b508541a
17. Petrenko, T., Kossmann, S., Neese, F.: Efficient time-dependent density functional theory approximations for hybrid density functionals: analytical gradients and parallelization. J. Chem. Phys. **134**(5), 054116 (2011). https://doi.org/10.1063/1.3533441

18. Perdew, J.P., Burke, K., Ernzerhof, M.: Generalized gradient approximation made simple. Phys. Rev. Lett. **77**(18), 3865–3868 (1996). https://doi.org/10.1103/PhysRevLett.77.3865
19. Perdew, J.P., Ernzerhof, M., Burke, K.: Rationale for mixing exact exchange with density functional approximations. J. Chem. Phys. **105**(22), 9982–9985 (1996). https://doi.org/10.1063/1.472933
20. Adamo, C., Barone, V.: Toward reliable density functional methods without adjustable parameters: The PBE0 model. J. Chem. Phys. **110**(13), 6158–6170 (1999). https://doi.org/10.1063/1.478522
21. Hess, B.A.: Applicability of the no-pair equation with free-particle projection operators to atomic and molecular structure calculations. Phys. Rev. A **32**(2), 756–763 (1985). https://doi.org/10.1103/PhysRevA.32.756
22. Hess, B.A.: Relativistic electronic-structure calculations employing a two-component no-pair formalism with external-field projection operators. Phys. Rev. A **33**(6), 3742–3748 (1986). https://doi.org/10.1103/PhysRevA.33.3742
23. Jansen, G., Hess, B.A.: Revision of the Douglas-Kroll transformation. Phys. Rev. A **39**(11), 6016–6017 (1989). https://doi.org/10.1103/PhysRevA.39.6016
24. Pantazis, D.A., Chen, X.-Y., Landis, C.R., Neese, F.: All-electron scalar relativistic basis sets for third-row transition metal atoms. J. Chem. Theory Comput. **4**(6), 908–919 (2008). https://doi.org/10.1021/ct800047t
25. Izsák, R., Neese, F.: An overlap fitted chain of spheres exchange method. J. Chem. Phys. **135**(14), 144105 (2011). https://doi.org/10.1063/1.3646921
26. Neese, F., Wennmohs, F., Hansen, A., Becker, U.: Efficient, approximate and parallel Hartree-Fock and hybrid DFT calculations. A 'chain-of-spheres' algorithm for the Hartree-Fock exchange. Chem. Phys. **356**(1), 98–109 (2009). https://doi.org/10.1016/j.chemphys.2008.10.036
27. Chang, S.-Y., Molleta, L.B., Booth, S.G., Uehara, A., Mosselmans, J.F.W., Ignatyev, K., Dryfe, R.A.W., Schroeder, S.L.M.: Automated analysis of XANES: a feasibility study of Au reference compounds. J. Phys: Conf. Ser. **712**(1), 012070 (2016). https://doi.org/10.1088/1742-6596/712/1/012070
28. Rehr, J.J., Albers, R.C.: Theoretical approaches to X-ray absorption fine structure. Rev. Mod. Phys. **72**(3), 621–654 (2000). https://doi.org/10.1103/RevModPhys.72.621
29. Westre, T.E., Kennepohl, P., DeWitt, J.G., Hedman, B., Hodgson, K.O., Solomon, E.I.: A multiplet analysis of Fe K-edge 1 s $\rightarrow$ 3d pre-edge features of iron complexes. J. Am. Chem. Soc. **119**(27), 6297–6314 (1997). https://doi.org/10.1021/ja964352a
30. van der Veen, R.M., Kas, J.J., Milne, C.J., Pham, V.-T., Nahhas, A., El Lima, F.A., Vithanage, D.A., Rehr, J.J., Abela, R., Chergui, M.: L-edge XANES analysis of photoexcited metal complexes in solution. Phys. Chem. Chem. Phys. **12**(21), 5551 (2010). https://doi.org/10.1039/b927033g
31. Alperovich, I., Smolentsev, G., Moonshiram, D., Jurss, J.W., Concepcion, J.J., Meyer, T.J., Soldatov, A., Pushkar, Y.: Understanding the electronic structure of 4d metal complexes: from molecular spinors to L-edge spectra of a di-Ru catalyst. J. Am. Chem. Soc. **133**(39), 15786–15794 (2011). https://doi.org/10.1021/ja207409q
32. Larabee, J.L., Hocker, J.R., Hanas, J.S.: Mechanisms of aurothiomalate-Cys2His2 zinc finger interactions. Chem. Res. Toxicol. **18**(12), 1943–1954 (2005). https://doi.org/10.1021/tx0501435
33. Du, Z., de Paiva, R.E.F., Nelson, K., Farrell, N.P.: Diversity in gold finger structure elucidated by traveling-wave ion mobility mass spectrometry. Angew. Chemie Int. Ed. **56**(16), 4464–4467 (2017). https://doi.org/10.1002/anie.201612494
34. DeBeer George, S., Petrenko, T., Neese, F.: Prediction of iron K-edge absorption spectra using time-dependent density functional theory. J. Phys. Chem. A **112**(50), 12936–12943 (2008). https://doi.org/10.1021/jp803174m
35. Lima, F.A., Bjornsson, R., Weyhermüller, T., Chandrasekaran, P., Glatzel, P., Neese, F., DeBeer, S., Shah, V.K., Konig, C., van Bokhoven, J.A., et al.: High-resolution molybdenum K-edge X-ray absorption spectroscopy analyzed with time-dependent density functional theory. Phys. Chem. Chem. Phys. **15**(48), 20911 (2013). https://doi.org/10.1039/c3cp53133c

36. Bjornsson, R., Lima, F.A., Spatzal, T., Weyhermüller, T., Glatzel, P., Bill, E., Einsle, O., Neese, F., DeBeer, S., Hoffman, B.M.: Identification of a spin-coupled Mo(III) in the nitrogenase iron–molybdenum cofactor. Chem. Sci. **5**(8), 3096–3103 (2014). https://doi.org/10.1039/C4SC00337C
37. Kowalska, J.K., Hahn, A.W., Albers, A., Schiewer, C.E., Bjornsson, R., Lima, F.A., Meyer, F., DeBeer, S.: X-ray absorption and emission spectroscopic studies of $[L_2Fe_2S_2]^n$ model complexes: implications for the experimental evaluation of redox states in iron-sulfur clusters. Inorg. Chem. **55**(9), 4485–4497 (2016). https://doi.org/10.1021/acs.inorgchem.6b00295
38. Hugenbruch, S., Shafaat, H.S., Krämer, T., Delgado-Jaime, M.U., Weber, K., Neese, F., Lubitz, W., DeBeer, S., Weng, T.-C., Zwart, P.H., et al.: In search of metal hydrides: an X-ray absorption and emission study of [NiFe] hydrogenase model complexes. Phys. Chem. Chem. Phys. **18**(16), 10688–10699 (2016). https://doi.org/10.1039/C5CP07293J
39. Roemelt, M., Beckwith, M.A., Duboc, C., Collomb, M.-N., Neese, F., DeBeer, S.: Manganese K-edge X-Ray absorption spectroscopy as a probe of the metal-ligand interactions in coordination compounds. Inorg. Chem. **51**(1), 680–687 (2012). https://doi.org/10.1021/ic202229b
40. Rees, J.A., Wandzilak, A., Maganas, D., Wurster, N.I.C., Hugenbruch, S., Kowalska, J.K., Pollock, C.J., Lima, F.A., Finkelstein, K.D., DeBeer, S.: Experimental and theoretical correlations between vanadium K-edge X-ray absorption and Kβ emission spectra. JBIC, J. Biol. Inorg. Chem. **21**(5–6), 793–805 (2016). https://doi.org/10.1007/s00775-016-1358-7
41. Bjornsson, R., Delgado-Jaime, M.U., Lima, F.A., Sippel, D., Schlesier, J., Weyhermüller, T., Einsle, O., Neese, F., DeBeer, S., Weyherm¸ller, T., et al.: Molybdenum L-edge XAS spectra of MoFe nitrogenase **641**(1), 65–71 (2015). https://doi.org/10.1002/zaac.201400446
42. Posewitz, M.C., Wilcox, D.E.: Properties of the Sp1 zinc finger 3 peptide: coordination chemistry, redox reactions, and metal binding competition with metallothionein. Chem. Res. Toxicol. **8**(8), 1020–1028 (1995). https://doi.org/10.1021/tx00050a005
43. Mély, Y., De Rocquigny, H., Morellet, N., Roques, B.P., Gérard, D.: Zinc binding to the HIV-1 nucleocapsid protein: a thermodynamic investigation by fluorescence spectroscopy. Biochemistry **35**(16), 5175–5182 (1996). https://doi.org/10.1021/bi952587d
44. Messori, L., Balerna, A., Ascone, I., Castellano, C., Gabbiani, C., Casini, A., Marchioni, C., Jaouen, G., Congiu Castellano, A.: X-ray absorption spectroscopy studies of the adducts formed between cytotoxic gold compounds and two major serum proteins. J. Biol. Inorg. Chem. (JBIC) **16**(3), 491–499 (2011). https://doi.org/10.1007/s00775-010-0748-5

Chapter 4
Probing Cells: Evaluating Cytotoxicity

4.1 Introduction

Auranofin is an FDA approved drug for the treatment of rheumatoid arthritis. In a repurposing effort, it has been extensively tested as an anticancer drug throughout the past decade. Regardless of the potent cytotoxicity observed for auranofin, there are many reports on lack of selectivity. Early mechanistic studies evidenced that auranofin and the precursor [AuCl(Et_3P)] affect mitochondrial function [1], and these results have been recently interpreted as induction of programmed cell death rather than inducing growth arrest of cycling cells [2]. Auranofin was shown to induce mitochondrial swelling and loss of membrane potential, related to mitochondrial membrane permeability transition (MPT) [3, 4]. The chemical state of the thiols present in mitochondria are important for controlling MPT and the Trx/TrxR system was expected to play a role. Auranofin was shown to be a more potent inhibitor of mitochondrial TrxR than [AuCl(Et_3P)] [5] but [AuCl(Et_3P)] is still more cytotoxic. Differences on the inhibition of TrxR point to different mechanisms of cytotoxicity between the two compounds.

Inspired by the structure of auranofin, and aiming to improve the cytotoxic selectivity, we decided to evaluate the cytotoxicity of the Au(I)-phosphine series of compounds (Fig. 4.1). The correlation between chemical structure and reactivity of the compounds with model biomolecules such as *N*-Ac-Cys and ZnFs was deeply discussed in Part I—Chaps. 1 and 3. Furthermore, the very same aspects that govern reactivity in the molecular level, such as the basicity and bulkiness of the phosphine ligand, σ-donating properties and lability of the co-ligands L (L = Cl or 4-dimethylaminopyridine, dmap) and overall charge of the compounds (neutral vs. cationic) are also expected to play a role in cytotoxic response, establishing an interesting set of parameters that can be correlated.

R. E. Ferraz de Paiva, *Gold(I,III) Complexes Designed for Selective Targeting and Inhibition of Zinc Finger Proteins*, Springer Theses,
https://doi.org/10.1007/978-3-030-00853-6_4

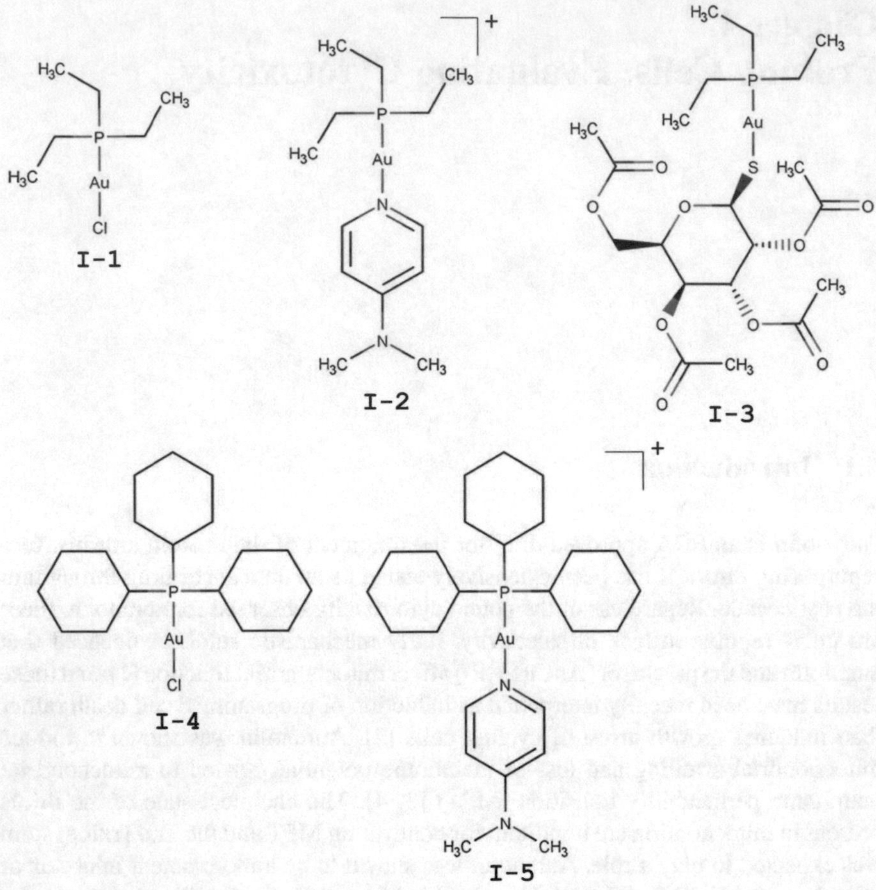

Fig. 4.1 Designed gold(I) compounds (**I-1**, **I-2**, **I-4** and **I-5**). Auranofin (**I-3**) was studied within the Et_3P series due to structural similarities

Regarding the our selection of model cell lines, auranofin went through phase II clinical trials for treating Chronic Lymphocytic Leukemia [6]. For that reason, the present chapter focuses on evaluating the cytotoxic effects of the Au(I)-phosphine series of compounds on T lymphocity tumorigenic cell line (CEM). The non-tumorigenic cell line HUVEC (human umbilical vein endothelial cells) was studied for comparison to determine selectivity. To further understand the factors governing cytotoxicity, additional assays were used, such as cell uptake of Au, evaluation of cell cycle arrest and determination of apoptosis-related protein expression profile.

4.2 Experimental

4.2.1 Materials

Compounds **I-1**, **I-2**, **I-3** and **I-4** were synthesized as described in Part I—Chap. 1. Auranofin was acquired from Sigma-Aldrich.

4.2.2 Cytotoxicity Assay

Human umbilical vein epithelial cells (HUVECs) were seeded in 96-well plates (5000 cells/well) containing 100 μL of Vasculife Endothelial Media (Lifeline), and allowed to attach overnight at 37 °C in 5% CO_2. The cells were treated with various concentrations of compounds **I-1** to **I-5** in sets containing 4 replicates per concentration. After 72 h, the media was removed and 0.5 mg mL^{-1} MTT (3-(4,5-dimethylthiazol-2-yl)-2,5-diphenyltetrazolium bromide) (Sigma) was added to each well and incubated for 3 h at 37 °C. The MTT reagent was removed and 100 μL of dmso were added to each well. Spectrophotometric readings were determined at 570 nm using a Synergy H1 microplate reader (BioTek). Percent cell survival was determined as treated/untreated controls ×100. The T lymphoblast cell line, CCRF-CEM, was seeded in 96-well plates (5000 cells/well) containing 100 μL of RPMI media. The cells were treated the same as above and assayed using the Cell Counting Kit 8 (Dojindo) according to the manufacturer's instructions.

4.2.3 Cellular Morphology

Following the same experimental conditions described for the cytotoxicity assay, but in a larger scale, 1×10^5 CEM cells were seeded in 100 mm dishes containing RPMI media and treated with IC_{25} and IC_{75} concentrations of the Au(I)-phosphine compounds Cells were then incubated at 37 °C (TC incubator) for 24 h. Images were acquired and processed using an Olympus IX70 inverted light microscope coupled with CellSens digital imaging software.

4.2.4 ICP-MS Analysis of Au Cellular Accumulation

Following the same experimental conditions described for the cytotoxicity assay, but in a larger scale, 1×10^6 CEM cells were seeded in 100 mm dishes containing RPMI media and treated with 0.18 μM of each compound for 10 min, 1.5 h or 3 h. Cells were harvested and washed twice with PBS. Cell pellets were digested overnight using

1 mL of conc. HNO_3 and diluted to 2 mL with water. Quantitation of Au content for each sample was analyzed using the Varian 820-MS ICP Mass Spectrometer. An 8-point standard curve (ranging from 100 to 0.7813 ppb) was generated for each experiment.

4.2.5 Cell Cycle Analysis Using Flow Cytometry

1×10^6 CEM cells were seeded in 100 mm dishes containing RPMI media. Cells were treated with IC_{75} concentrations of each compound and incubated for 6 or 24 h at 37 °C. Both floating and attached cells were harvested, washed, and resuspended in 1 mL of propidium iodide solution (3.8 mmol L^{-1} sodium citrate; 0.05 mg/mL propidium iodide; 0.1% Triton X-100) with added RNase B (7 Kunitz units/mL). After staining, the solutions were passed through 35 μm filters. The samples (20,000 events each) were immediately analyzed at 670 nm by flow cytometry using a BD FACSCanto II flow cytometer and FACSDiva software.

4.2.6 Antibody Array Analysis of Phospho-Protein Expression

The PathScan Stress and Apoptosis Signaling Antibody Array Kit (Cell Signaling) was used according to the manufacturer's instructions. Briefly, 2×10^6 CEM cells were seeded in two 100 mm dishes per sample and allowed to attach overnight. The cells were treated with IC_{75} concentrations of each compound for 3, 6 and 12 h time points. The cells were washed, harvested on ice, and lysed in the presence of protease and phosphatase inhibitors. The amount of protein for each sample was quantified using the Bradford assay. Blocking buffer was added to each array for 15 min at room temperature, removed and replaced with 75 μg of protein overnight at 4 °C. After removal of the protein, arrays were washed, and detection antibody cocktail supplied with the kit was added for 1 h at room temperature. The arrays were washed and incubated for 30 min with HRP-linked Streptavidin solution at room temperature. Finally, the slide was covered with lumiGLO/Peroxide reagent and the chemiluminescent array images were captured on film following 3–5 s exposure times. The density of the spots was quantified using the Protein Array Analyzer by ImageJ and normalized to the α-tubulin values for each sample. The difference in protein expression between the untreated control and treated samples is expressed as fold-change (normalized treated sample/normalized untreated control sample).

4.3 Results and Discussion

4.3.1 Cytotoxicity

Figure 4.2 shows the growth inhibition profile of CEMs and HUVECs treated with the five Au(I)-phosphine compounds. An interesting selectivity was observed towards the tumorigenic cell line, as observed by the comparison of IC_{50} (Table 4.1). The IC_{50} values obtained were always lower for CEMs than for HUVECs for every newly designed compound in the series (**I-1**, **I-2**, **I-4** and **I-5**), indicating the selectivity of the Au(I)-phosphine compounds towards the tumorigenic cell line over the normal one. Auranofin, on the other hand, had no selectivity between the two cell lines and was marginally more cytotoxic against the non-tumorigenic HUVEC cell line.

The compound $[Au(dmap)(Et_3P)]^+$ presented a>57 selectivity index towards the tumorigenic CEM cell line. The replacement of Cl by dmap dramatically decreases the reactivity rate of the Au(I) compound when targeting model biomolecules (N-acetyl-L-Cys) and also zinc fingers (HIV-1 nucleocapsid protein NCp7), in terms of the temporal evolution of the aurated species identified by MS (See Part1—Chap. 1).

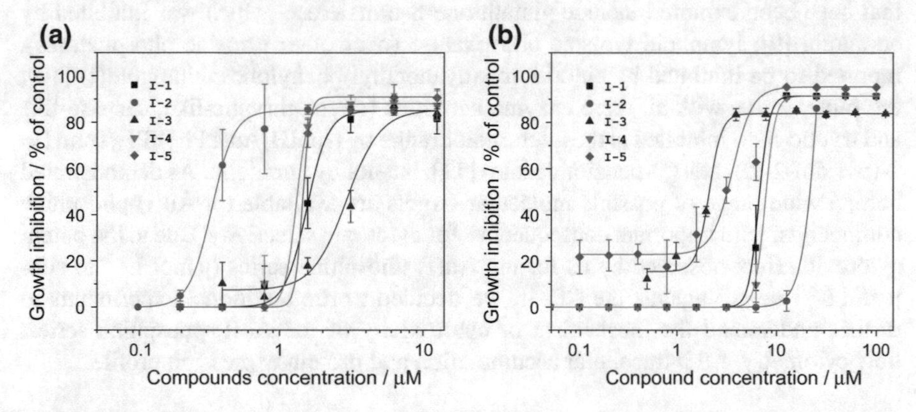

Fig. 4.2 Growth inhibition profile of compounds of **a** CEM and **b** HUVEC cell lines exposed to compounds **I-1** $[AuCl(Et_3P)]$, **I-2** $[Au(dmap)(Et_3P)]^+$, **I-3** Auranofin, **I-4** $[AuCl(Cy_3P)]$ and **I-5** $[Au(dmap)(Cy_3P)]^+$

Table 4.1 Summary of the profile of cytotoxicity of the Au(I)-phosphine compounds

Compound	IC_{50} / µmol L^{-1}	
	HUVEC	CEM
$[AuCl(Et_3P)]$	6.77 ± 0.15	1.54 ± 0.06
$[Au(dmap)(Et_3P)]^+$	18.44 ± 0.92	0.32 ± 0.01
Auranofin	2.13 ± 0.08	2.91 ± 0.48
$[AuCl(Cy_3P)]$	8.39 ± 0.34	1.29 ± 0.08
$[Au(dmap)(Cy_3P)]^+$	3.59 ± 0.39	1.35 ± 0.05

When comparing the ancillary phosphines, Et_3P-containing compounds (**I-1** and **I-2**) seem to be in general less reactive than Cy_3P-containing compounds (**I-4** and **I-5**). Slower reaction rate is often translated into higher metalation selectivity. Combining the effects of the phosphine and ligand L, $[Au(dmap)(Et_3P)]^+$ was the least reactive compound in the series when targeting NCp7 ZnF2, as characterized by CD, MS and ^{31}P NMR (See Part I—Chap. 1). The unique profile of cytotoxicity observed for **I-2** is translated from a unique chemical reactivity within the Au(I)-phosphine series studied here.

Figure 4.6 (see Appendix) shows the morphology of CEM cells untreated and treated with compounds **I-3**. Untreated CEMs have a distinguished feature of forming clusters of cells, which are disrupted upon treatment with the Au(I)-phosphine compounds. A larger amount of cellular debris and non-intact cells can be observed for treatment with compound $[AuCl(Et_3P)]$. Treatment with compound **I-2** led to the larger number of intact cells, while treatment with auranofin induced aggregation in the cytoplasm but most cells remained with intact cellular membranes.

Regarding molecular targets, $[AuCl(Et_3P)]$ was shown to bind to DNA [7], as opposed to auranofin. A series of Au(I)-triethylphosphine, including auranofin, was shown to inhibit cytosolic and mitochondrial thioredoxin reductase [8], the most recurring target in the literature for this class of compounds. Some other targets that have been explored include glutathione-S-transferase, which was inhibited by auranofin [9]; lymphoid tyrosine phosphatase (over other tyrosine phosphatases), reported to be inhibited by chlorido(bis(cyanoethyl)phenylphosphine)gold(I) [10]; the proteasome, with all three enzyme activities (chymothryptic-like, caspase-like and tryptic-like) inhibited in the micromolar range by $[(pbiH)Au(PPh_3)]PF_6$ (pbiH = 2-(pyridin-2-yl)-1H-1,3-benzimidazole) [11], but not by auranofin. As demonstrated here, a wide range of possible molecular targets are available for Au(I)-phosphine compounds, with important consequences for cytotoxic selectivity. Due to the potent cytotoxic effect observed by us for the Au(I)-phosphine series (μmol L^{-1} to sub-μmol L^{-1} range) against the CEMs, we decided to run additional experiments to further understand the mechanism of cytotoxicity of the Au(I)-phosphine series: flow cytometry, Au intracellular accumulation and protein expression profile.

4.3.2 *Flow Cytometry*

Flow cytometry was used to evaluate cellular integrity and to assess changes in CEM's cell cycle induced by treatment with Au(I)-phosphine compounds. It was interesting to observe that typical incubation times (24 h) for CEMs exposed to the Au(I)-phosphine compounds led to high apoptotic activity (Fig. 4.3a). The apoptosis-inducing behavior is clearly observed at the sub-G1 region. Cells treated with compound **I-5** ($[Au(dmap)(Cy_3P)]^+$) were the least affected, but still with a high population in the sub-G1 region when compared to control. The experimental conditions were further tuned in order to get the details lost at the highly cytotoxic treatment regimen. CEMs exposed to IC_{75} concentrations but at reduced incubation times

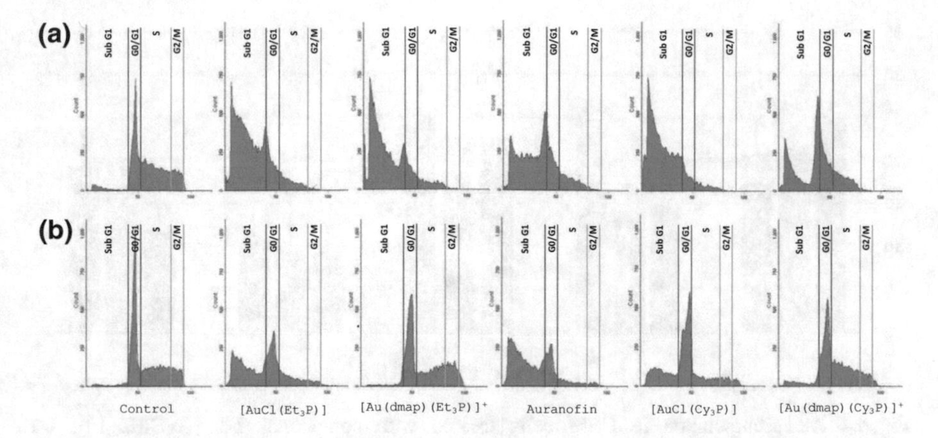

Fig. 4.3 Flow cytometry profile of CEM cells treated with IC_{75} concentrations for **a** 24 h, highlighting the highly apoptotic response caused by the Au(I)-phosphine series and **b** for 6 h, showing differences within the series. Cell counts corresponding to the Sub-G1 region are marked in red, the G0/G1 region in blue and the G2/M region in yellow

gave more information regarding the mechanism of action of the tested compounds (Fig. 4.3b). Auranofin and compound **I-1** ([AuCl(Et_3P)]) had a similar apoptotic effect, with a high population of dead cells/debris. Both Cy_3P containing compounds (**I-4** and **I-5**) induced an accumulation in the G1/S transition when compared to control. Cells treated with compound [Au(dmap)(Et_3P)]$^+$ (**I-2**) presented a particularly distinguished behavior, with reduced G1 population and a slight accumulation in G2. This behavior is indicative of a different mechanism of cell death, since the G2 arrest is often a result of DNA damage or mitotic catastrophe. Flow cytometry data obtained for treatments with IC_{25} for 6 h can be found in Fig. 4.7 (See Appendix).

As a comparison, previous studies indicated that B16 melanoma cells treated with auranofin in the range 0.0125–0.2 μmol L^{-1} for 24 h led to 90% or greater cell survival, with no changes from the control experiment observed in the DNA distribution histograms. Treatments with 0.2 μmol L^{-1} and higher concentrations led to extensive cellular lysis. Auranofin showed no selectivity against cycling versus noncycling cells [12].

4.3.3 Cellular Uptake of Au

Knowing the cellular accumulation is important for drug design and further development, since ineffectiveness is often related to low cellular uptake. A metallodrug facilitates the quantification of drug incorporated into cells, as the metal can be directly quantified by means of instrumental techniques such as inductively coupled plasma mass spectrometry (ICP-MS). Some trends can be observed analyzing the Au intracellular accumulation profile (Fig. 4.4), which can be corrolated with structural

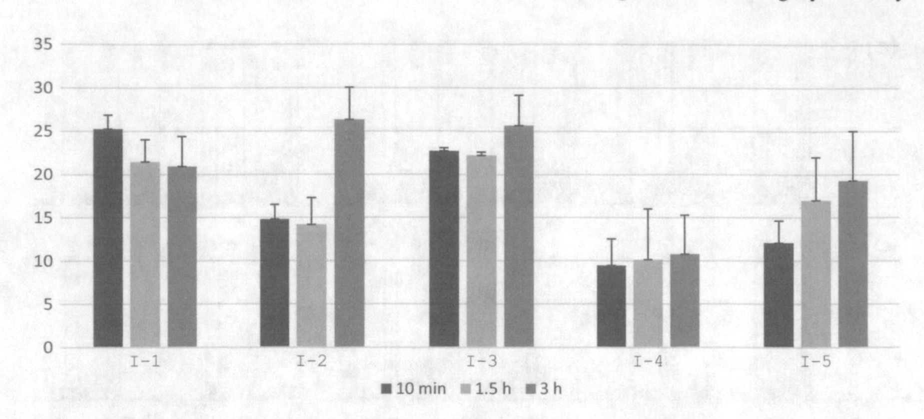

Fig. 4.4 Au accumulation in CEM cells treated with compounds **I-1** $[AuCl(Et_3P)]$, **I-2** $[Au(dmap)(Et_3P)]^+$, **I-3** Auranofin, **I-4** $[AuCl(Cy_3P)]$ and **I-5** $[Au(dmap)(Cy_3P)]^+$

properties of the Au(I) compounds. The size of the phosphine ligands has the first and most noticeable effect. Et_3P-containing compounds (**I-1**, **I-2** and **I-3**) have a higher final accumulation than Cy_3P-containing compounds (**I-4** and **I-5**). Figure 4.4 shows that the same final Au content, considering the standard deviation, was observed for CEMs treated with compounds **I-1** and **I-3**. A clear temporal evolution of the Au uptake by CEM cells can be observed for the cationic compounds (**I-2** and **I-5**). On the other hand the neutral compounds $[AuCl(Et_3P)]$, auranofin and $[AuCl(Cy_3P)]$ had an inconclusive temporal evolution behavior, with a constant Au incorporation (within the standard deviations) throughout the experiment.

The interaction of $[AuCl(Et_3P)]$ and auranofin with RAW 264.7 macrophages has been described in the literature [13]. The interaction of the compounds with the target cell line was based on ^{195}Au-labled compounds. A higher Au uptake was observed in RAW 264.7 cells treated with $[AuCl(Et_3P)]$ than with auranofin. The time course of the cellular association of the ^{195}Au-labled compounds with RAW 246.7 cells was also studied. The ^{195}Au content was evaluated for up to 60 min. The total ^{195}Au content increased in the first 20 min and then started to drop, as did the total number of cells (cytotoxic effect). The incorporation of Au from $[AuCl(Et_3P)]$ was higher than from auranofin [13].

4.3.4 Protein Expression Profile

The activation of 19 cell stress and apoptosis-related proteins in CEMS cells in response to IC_{75} treatments with auranofin, $[AuCl(Et_3P)]$ and $[Au(dmap)(Et_3P)]^+$ were compared using phospho-antibody array kits (Fig. 4.5a). The Pathscan Stress and Apoptosis Signaling antibody array kit with chemiluminescent readouts makes it easy to simultaneously detect the cleavage of poly (ADP-ribose) polymerase (PARP) (Asp214), caspase-3 (Asp175) and caspase-7 (Asp198); phosphorylation of extra-

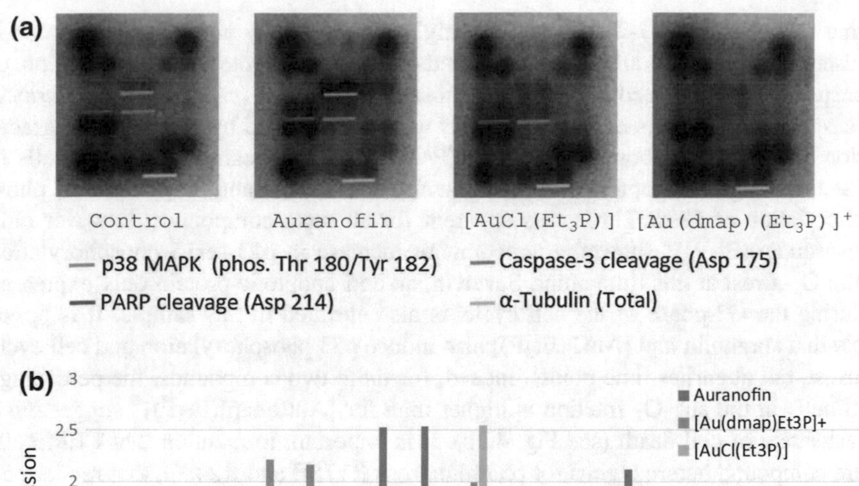

Fig. 4.5 **a** CEM cells either untreated (control) or treated with Au(I)-phosphine compounds **I-1** to **I-3**. Cell lysates were analyzed with the PathScan stress and apoptosis signaling antibody array kit. **b** Densitometric protein profile of CEM cells treated with compounds **I-1** to **I-3**

cellular signal-regulated kinase (ERK)-1/2 (Thr202/Tyr204), protein kinase B (Akt; Ser473), BCL-2-associated death promoter (Bad; Ser136), heat shock protein 27 (HSP27; Ser82), Smad2 (Ser465/467), p53 (Ser15), p38 mitogen activated protein kinase (p38 MAPK; Thr180/Tyr182), stress-activated protein kinases and Jun amino-terminal kinases (SAPK/JNK; Thr183/Tyr185), checkpoint kinase 1 (Chk1; Ser345), checkpoint kinase 2 (Chk2; Thr68), NF-κB inhibitor α (IκBα Ser32/36), eukaryotic translation initiation factor 2 subunit α (elF2; Ser51) and transforming growth factor-β-activated kinase 1 (TAK1) (Ser412) and determination of protein content of IκBα, survivin and α-tubulin. These signaling molecules represent a good starting point for probing the mechanism of cell death when studying a new cytotoxic agent. Quantification using densitometric analysis is also shown (Fig. 4.5b).

The Au(I) series containing triethylphosphine (compounds **I-1**, **I-2** and **I-3**) was selected because it demonstrated better selectivity towards the tumorigenic cell line and compound **I-2** exhibited the highest cytotoxicity among the compounds discussed here. For all three compounds, significant proteolytic degradation of caspase-3 was observed, an important substrate of intrinsic (caspase-9) and extrinsic (caspase-8) mediators of apoptosis. This was accompanied by proteolytic degradation of poly(ADP-ribose)-polymerase (PARP), a downstream indicator of cells in the final stage of apoptotic DNA fragmentation. DNA strand break-induced phosphorylation of Chk2 Thr68 was apparent for all three compounds, however only $[Au(dmap)(Et_3P)]^+$ showed a concomitant increase in p53 ser15 phosphorylation and G_2-arrest at this time point. Survivin, an anti-apoptotic protein only expressed during the G2-phase of the cell cycle, is also elevated in this sample. It is possible that auranofin and $[AuCl(Et_3P)]$ also induce p53 phosphorylation and cell cycle arrest, but at earlier time points. Indeed, for these two compounds, the percentage of cells in the sub-G_1 fraction is higher than for $[Au(dmap)(Et_3P)]^+$ suggesting a faster rate of cell death (see Fig. 4.3b). It is important to mention that CEM cells are compound heterozygous for p53 mutations R175H and R248Q, that render p53 unable to activate typical response genes and to mediate apoptosis [14]. As observed in Fig. 4.4, although the final concentration of gold per cell is similar for the 3 compounds evaluated, $[AuCl(Et_3P)]$ and auranofin had a higher accumulation at earlier time points, while $[Au(dmap)(Et_3P)]^+$ took a longer time to reach the final concentration observed. These observations are suggestive of an alternative mechanism of cell death for $[Au(dmap)(Et_3P)]^+$, a direct consequence of the slower chemical reactivity achieved by replacing the labile chloride by dmap.

CEM cells treated with auranofin showed increased phosphorylation of p38 MAPK (Thr180/Tyr182) and SAPK/JNK (Thr183/Tyr185), as previously reported for HL-60 leukemia cells [15]. This result is consistent with the well-documented cellular response to the increase of reactive oxygen species due to inhibition of cytosolic and mitochondrial thioredoxin reductase (TrxR) [16, 17]. Furthermore, $[Au(dmap)(Et_3P)]^+$ showed elevated levels of Smad2/3 (Ser465/467) phosphorylation when compared to the untreated control. Auranofin, in fact, showed the opposite result. Although we cannot be certain Smad2/3 is not activated at earlier or later time points in this system, to our knowledge an increase in Smad2/3 phosphorylation have never been reported for auranofin. Smad2 and Smad3 become phosphorylated at their carboxyl termini by the receptor kinase TGF-β Receptor I following stimulation by TGF-β. Recent reports correlate activation of TGF-β with induction of apoptosis [18, 19]. Latent TGF-β acts as an extracellular sensor of oxidative stress. Release of active TGF-β from the latent complex binding protein (LTBP) and latency associated peptide (LAP) occurs in response to a variety of agents such as heat, acidic pH, chaotropic agents, and reactive oxygen species.

ERK1 and ERK2 are protein-serine/threonine kinases that participate in the Ras-Raf-MEK-ERK cascade. This cascade is responsible for regulating a large variety of cellular processes, mainly proliferation and survival but also including adhesion, cell cycle progression, migration and differentiation [20]. The ERK1/2 catalyzed phosphorylation of nuclear transcription factors requires the translocation of ERK1/2 into the nucleus by active and passive processes involving the nuclear pore. These transcription factors participate in the immediate early gene response. Auranofin and $[AuCl(Et_3P)]$ led to ERK1/2 phosphorylation levels similar to that of the untreated control. On the other hand, increased ERK1/2 phosphorylation was observed for CEM cells treated with $[Au(dmap)(Et_3P)]^+$. Although ERK1/2 phosphorylation generally promotes cell survival, it has been shown that under certain circumstances of DNA damage stimuli, ERK1/2 can have pro-apoptotic functions. For example, ERK promotes p53 stability by phosphorylation on Ser15 [21, 22], leading to its accumulation by inhibition of the association with Mdm2 [23]. CEM cells treated with $[Au(dmap)(Et_3P)]^+$ had both ERK 1/2 and p53 phosphorylation upregulated. Furthermore, it has been hypothesized that the main cause of ERK sustained activation is the presence of ROS [24].

Finally, HSP27 is a mediator of cell stress, conferring resistance to adverse conditions, being activated by phosphorylation at Ser82. CEM cells treated with $[Au(dmap)(Et^3P)]^+$ had a significant increase in HSP27 phosphorylation, which is not observed for treatment with $[AuCl(Et_3P)]$. A significant HSP27 phosphorylation induction was also recently observed when HT-29 cells were treated with organometallic Au(I)(triphenylphosphine) compounds containing alkynyl and N-heterocyclic carbene ligands [25, 26]. HSP27 activation, along with eIF2, is consistent with a mechanism based on extracellular oxidative stress.

4.4 Conclusions

A series of Au(I)-phosphine compounds based on the Et_3P and on the bulkier Cy_3P ligands was evaluated in terms of cytotoxicity. Two cell lines were selected, a T lymphocyte tumorigenic cell line (CEM) and the non-tumorigenic the Human Umbilical Vein Endothelial cells (HUVEC). IC_{50} values obtained using the MTT protocol were found in lower μM range, or even sub-μM for the compound $[Au(dmap)(Et_3P)]^+$. All compounds evaluated here, besides auranofin, demonstrated selectivity towards the CEM cell line. The highest selectivity (>57×) was found for $[Au(dmap)(Et_3P)]^+$. High apoptotic activity was observed on CEM cells treated with each compound tested here based on flow cytometry data. Under typical exposure time (24 h), a high population of CEMs in the sub-G1 region was observed, relative to the untreated control, caused by treatment with the Au(I)-phosphine compound. Reducing the exposure time to just 6 h, CEMs treated with $[Au(dmap)(Et_3P)]^+$ showed a very particular behavior within the series, with arrest in G2. The CEM cells treated with Et_3P-containing compounds were further evaluated by protein expression profile. The compound $[Au(dmap)(Et_3P)]^+$ was shown to cause a wide range of cellular responses that trigger apoptosis (based for example on activation of Chk2, HSP27, sustained activation of ERK 1/2), as opposed to auranofin that mainly activates p38 MAPK. The sub-μM cytotoxicity observed over CEM cells is a direct consequence of this multi-modal response. The mechanism of apoptosis induced in CEM cells by treatment with $[Au(dmap)(Et_3P)]^+$ is summarized in Scheme 4.1. For comparison, the overall mechanism of cytotoxicity caused by auranofin is given in Fig. 4.8 (See Appendix), which is in agreement with the apoptotic mechanism previously reported for HL-60 leukemia cells treated with auranofin [15].

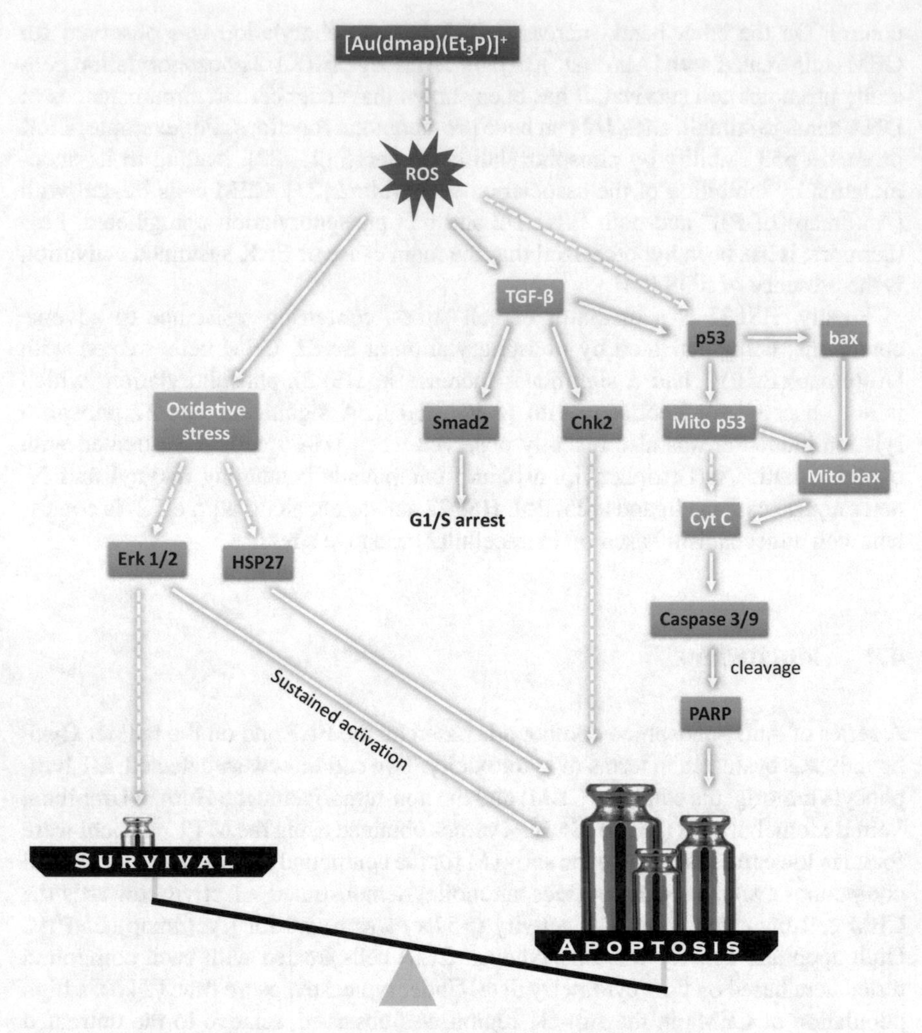

Scheme 4.1 Apoptotic mechanism proposed for $[Au(dmap)(Et_3P)]^+$. The proteins directly observed written in blue, while the expected/required intermediates are written in white

Appendix

Cellular Morphology

See Fig. 4.6.

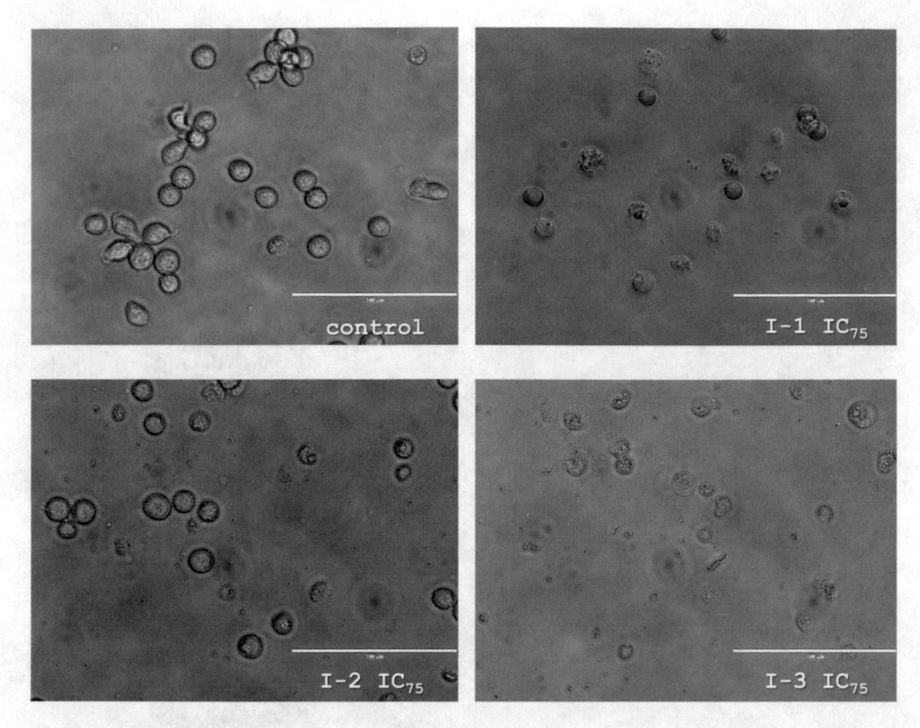

Fig. 4.6 Morphology of CEM cell line untreated and treated with IC_{75}-concentrations of compounds $[AuCl(Et_3P)]$ (**I-1**), $[Au(dmap)(Et_3P)]^+$ (**I-2**) and auranofin (**I-3**)

Flow Cytometry

See Fig. 4.7.

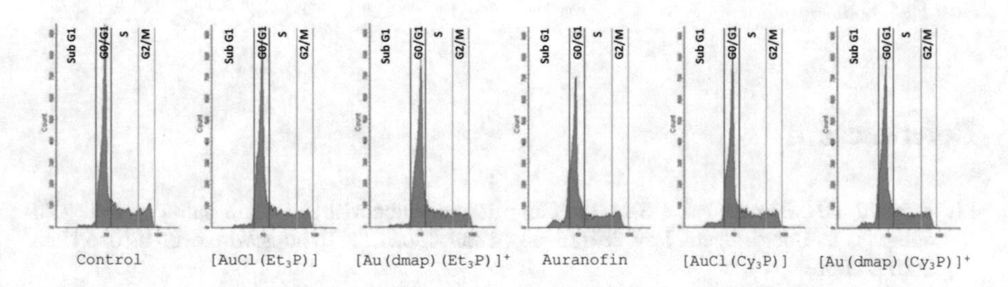

Fig. 4.7 Flow cytometry profile of CEM cell treated with IC_{25} concentrations for 6 h

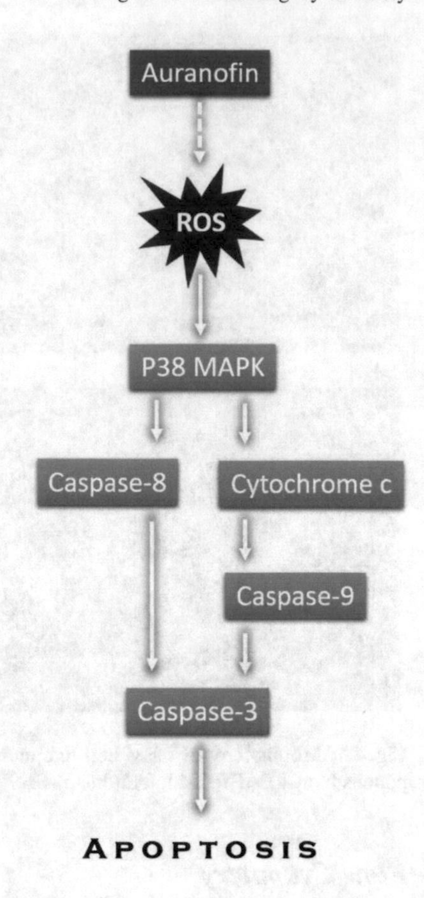

Fig. 4.8 Apoptotic mechanism caused by auranofin on CEM cells. The mechanism is consistent with previous results on HL-60 leukemia cells treated with auranofin [15]

Mechanism of Auranofin-Induced Apoptosis

See Fig. 4.8.

References

1. Barnard, P.J., Berners-Price, S.J.: Targeting the mitochondrial cell death pathway with gold compounds. Coord. Chem. Rev. **251**(13–14), 1889–1902 (2007). https://doi.org/10.1016/j.ccr.2007.04.006
2. McKeage, M.J.: Gold opens mitochondrial pathways to apoptosis. Br. J. Pharmacol. **136**(8), 1081–1082 (2002). https://doi.org/10.1038/sj.bjp.0704822
3. Rigobello, M.P., Scutari, G., Folda, A., Bindoli, A.: Mitochondrial thioredoxin reductase inhibition by gold(I) compounds and concurrent stimulation of permeability transition and release of cytochrome c. Biochem. Pharmacol. **67**(4), 689–696 (2004). https://doi.org/10.1016/j.bcp.2003.09.038

4. Rigobello, M.P., Scutari, G., Boscolo, R., Bindoli, A.: Induction of mitochondrial permeability transition by auranofin, a gold(I)-phosphine derivative. Br. J. Pharmacol. **136**(8), 1162–1168 (2002). https://doi.org/10.1038/sj.bjp.0704823
5. Rigobello, M.P., Callegaro, M.T., Barzon, E., Benetti, M., Bindoli, A.: Purification of Mitochondrial Thioredoxin Reductase and Its Involvement in the Redox Regulation of Membrane Permeability, vol. 24 (1998)
6. Phase I and II Study of Auranofin in Chronic Lymphocytic Leukemia (CLL)—Full Text View—ClinicalTrials.gov https://clinicaltrials.gov/ct2/show/nct01419691?term=auranofin+cll&rank=1
7. Mirabelli, C.K., Sung, C.-M., Zimmerman, J.P., Hill, D.T., Mong, S., Crooke, S.T.: Interactions of gold coordination complexes with DNA. Biochem. Pharmacol. **35**(9), 1427–1433 (1986). https://doi.org/10.1016/0006-2952(86)90106-1
8. Gandin, V., Fernandes, A.P., Rigobello, M.P., Dani, B., Sorrentino, F., Tisato, F., Björnstedt, M., Bindoli, A., Sturaro, A., Rella, R., et al.: Cancer cell death induced by phosphine gold(I) compounds targeting thioredoxin reductase. Biochem. Pharmacol. **79**(2), 90–101 (2010). https://doi.org/10.1016/j.bcp.2009.07.023
9. De Luca, A., Hartinger, C.G., Dyson, P.J., Lo Bello, M., Casini, A.: A new target for gold(I) compounds: glutathione-S-transferase inhibition by auranofin. J. Inorg. Biochem. **119**, 38–42 (2013). https://doi.org/10.1016/j.jinorgbio.2012.08.006
10. Karver, M.R., Krishnamurthy, D., Bottini, N., Barrios, A.M.: Gold(I) phosphine mediated selective inhibition of lymphoid tyrosine phosphatase. J. Inorg. Biochem. **104**(3), 268–273 (2010). https://doi.org/10.1016/j.jinorgbio.2009.12.012
11. Micale, N., Schirmeister, T., Ettari, R., Cinellu, M.A., Maiore, L., Serratrice, M., Gabbiani, C., Massai, L., Messori, L.: Selected cytotoxic gold compounds cause significant inhibition of 20S proteasome catalytic activities. J. Inorg. Biochem. **141**, 79–82 (2014). https://doi.org/10.1016/j.jinorgbio.2014.08.001
12. Mirabelli, C.K., Johnson, R.K., Sung, C., Faucette, L., Muirhead, K., Crooke, S.T.: Evaluation of the in vivo antitumor activity and in vitro cytotoxic properties of auranofin, a coordinated gold compound, in murine tumor models. Cancer Res. **45**, 32–39 (1985)
13. Snyder, R.M., Mirabelli, C.K., Crooke, S.T.: Cellular interactions of auranofin and a related gold complex with raw 264.7 macrophages. Biochem. Pharmacol. **36**(5), 647–654 (1987). https://doi.org/10.1016/0006-2952(87)90715-5
14. Eustermann, S., Videler, H., Yang, J.C., Cole, P.T., Gruszka, D., Veprintsev, D., Neuhaus, D.: The DNA-binding domain of human PARP-1 interacts with DNA single-strand breaks as a monomer through its second zinc finger. J. Mol. Biol. **407**(1), 149–170 (2011). https://doi.org/10.1016/j.jmb.2011.01.034
15. Park, S.-J., Kim, I.-S.: The role of p38 MAPK activation in auranofin-induced apoptosis of human promyelocytic leukaemia HL-60 cells. Br. J. Pharmacol. **146**(4), 506–513 (2005). https://doi.org/10.1038/sj.bjp.0706360
16. Omata, Y., Folan, M., Shaw, M., Messer, R.L., Lockwood, P.E., Hobbs, D., Bouillaguet, S., Sano, H., Lewis, J.B., Wataha, J.C.: Sublethal concentrations of diverse gold compounds inhibit mammalian cytosolic thioredoxin reductase (TrxR1). Toxicol. Vitr. **20**(6), 882–890 (2006). https://doi.org/10.1016/j.tiv.2006.01.012
17. Omata, Y., Lewis, J.B., Lockwood, P.E., Tseng, W.Y., Messer, R.L., Bouillaguet, S., Wataha, J.C.: Gold-induced reactive oxygen species (ROS) do not mediate suppression of monocytic mitochondrial or secretory function. Toxicol. In Vitro **20**(5), 625–633 (2006). https://doi.org/10.1016/j.tiv.2005.11.001
18. Solovyan, V.T., Keski-Oja, J.: Apoptosis of human endothelial cells is accompanied by proteolytic processing of latent TGF-β binding proteins and activation of TGF-β. Cell Death Differ. **12**(7), 815–826 (2005). https://doi.org/10.1038/sj.cdd.4401618
19. Richter, K., Konzack, A., Pihlajaniemi, T., Heljasvaara, R., Kietzmann, T.: Redox-fibrosis: impact of TGFβ1 on ROS generators, mediators and functional consequences. Redox Biol. **6**, 344–352 (2015). https://doi.org/10.1016/j.redox.2015.08.015

20. Roskoski, R.: ERK1/2 MAP kinases: structure, function, and regulation. Pharmacol. Res. **66**(2), 105–143 (2012). https://doi.org/10.1016/j.phrs.2012.04.005
21. Woessmann, W., Chen, X., Borkhardt, A.: Ras-mediated activation of ERK by cisplatin induces cell death independently of p53 in osteosarcoma and neuroblastoma cell lines. Cancer Chemother. Pharmacol. **50**(5), 397–404 (2002). https://doi.org/10.1007/s00280-002-0502-y
22. Liu, J., Mao, W., Ding, B., Liang, C.-S.: ERKs/p53 signal transduction pathway is involved in doxorubicin-induced apoptosis in H9c2 cells and cardiomyocytes. AJP Hear. Circ. Physiol. **295**(5), H1956–H1965 (2008). https://doi.org/10.1152/ajpheart.00407.2008
23. She, Q.B., Chen, N., Dong, Z.: ERKs and p38 kinase phosphorylate p53 protein at serine 15 in response to UV radiation. J. Biol. Chem. **275**(27), 20444–20449 (2000). https://doi.org/10.1074/jbc.M001020200
24. Cagnol, S., Chambard, J.-C.: ERK and cell death: mechanisms of ERK-induced cell death—apoptosis, autophagy and senescence. FEBS J. **277**(1), 2–21 (2010). https://doi.org/10.1111/j.1742-4658.2009.07366.x
25. Andermark, V., Göke, K., Kokoschka, M., Abu el Maaty, M.A., Lum, C.T., Zou, T., Sun, R.W.-Y., Aguiló, E., Oehninger, L., Rodríguez, L., et al.: Alkynyl gold(I) phosphane complexes: evaluation of structure–activity-relationships for the phosphane ligands, effects on key signaling proteins and preliminary in-vivo studies with a nanoformulated complex. J. Inorg. Biochem. **160**, 140–148 (2016). https://doi.org/10.1016/j.jinorgbio.2015.12.020
26. Holenya, P., Can, S., Rubbiani, R., Alborzinia, H., Jünger, A., Cheng, X., Ott, I., Wölfl, S.: Detailed analysis of pro-apoptotic signaling and metabolic adaptation triggered by a N-heterocyclic carbene–gold(I) complex. Metallomics **6**(9), 1591–1601 (2014). https://doi.org/10.1039/C4MT00075G

Part II
Au(III) Complexes

Chapter 5
Au(III) Series with κ^2C,N and κ^2N,N′ Ligands

5.1 Introduction

Gold(III) complexes are often seen as unstable in biological media, mainly due to the high content of sulfur-containing species that lead to redox processes. On the other hand, stabilizing Au(III) represents an interesting synthetic challenge, given the wide variety of possible strategies available. For that reason, we decided to explore the stabilization of Au(III) using the N^C donor 2-benzylpyridine (bnpy) ligand and compare it to the classic bipy-like N^N coordination motif (Fig. 5.1).

Although presenting many stability-related advantages over the Au(N^N) motif, the Au(C^N) motif has been much less explored in the literature, with few examples reported [1–4]. Au(C^N) organometallic complexes have been studied mainly for anticancer applications. Parish introduced, in 1996, the complexes [AuX_2(damp)] (damp = o-$C_6H_4CH_2NMe_2$; X = Cl^-, OAc^- or X_2 = dithiocarbamate, malonate) [5]. The gold(III) complexes [Au(acetato)$_2$(damp)] and [Au(malonato)(damp)] exhibited selective cytotoxicity in vitro and showed in vivo antitumor activity against subcutaneously implanted xenografts derived from the HT1376 bladder and CH1 ovarian cell lines [6]. Au(III) complexes based on the C^N or C^N^N motifs (deprotonated C-donors) such as 2-(2-phenylpropan-2-yl)pyridine (pydmb-H) and 6-(2-phenylpropan-2-yl)-2,2′-bipyridine (bipydmb) ligands were studied by Messori and co-workers and exhibited remarkable redox and ligand replacement stability in biological media, even in the presence of relatively high concentrations of intracellular thiols [4, 7].

Regarding studies targeting biomolecules, the Au(C^N^N) compound containing the deprotonated 6-(1,1-dimethylbenzyl)-2,2′-bipyridine) as ligand demonstrated amino acid-level selectivity when reacted with hen egg white lysozyme. The reaction product (which had its crystal structure determined) had a Au(I) coordinated to a

---

Parts of this chapter has been reproduced with permission from Wiley. https://onlinelibrary.wiley.com/doi/abs/10.1002/anie.201803082

R. E. Ferraz de Paiva, *Gold(I,III) Complexes Designed for Selective Targeting and Inhibition of Zinc Finger Proteins*, Springer Theses,
https://doi.org/10.1007/978-3-030-00853-6_5

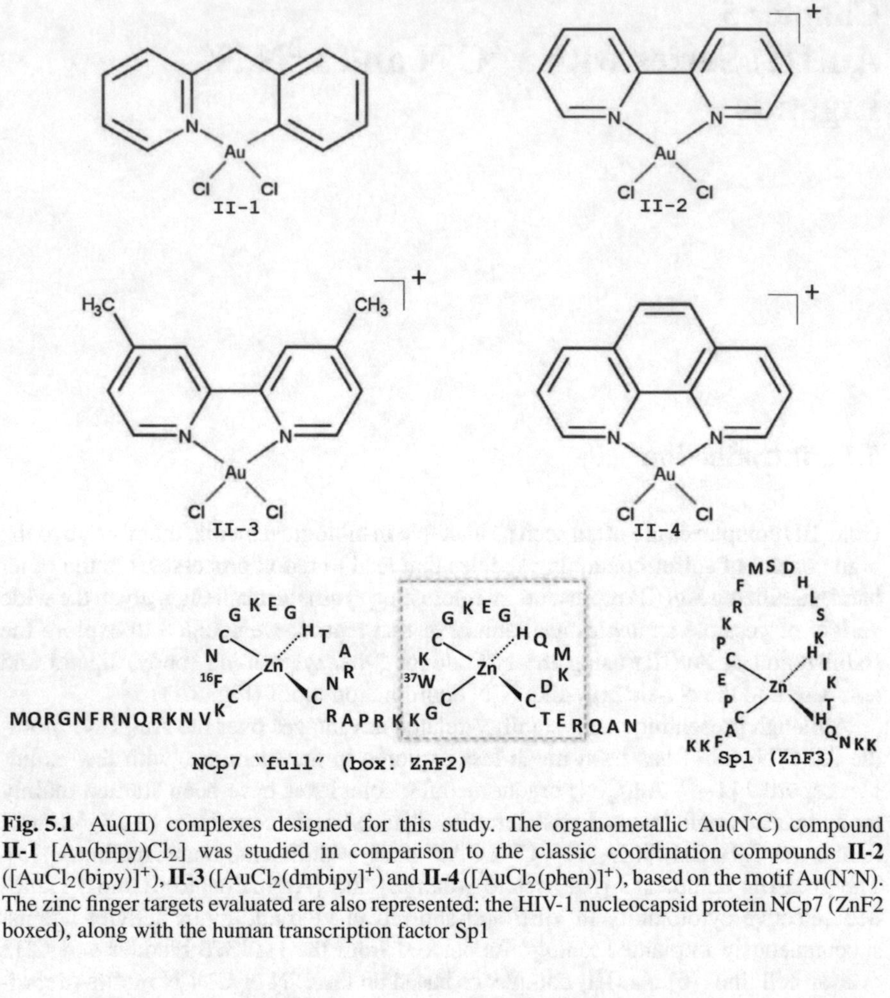

Fig. 5.1 Au(III) complexes designed for this study. The organometallic Au(N^C) compound **II-1** [$Au(bnpy)Cl_2$] was studied in comparison to the classic coordination compounds **II-2** ([$AuCl_2(bipy)]^+$), **II-3** ([$AuCl_2(dmbipy]^+$) and **II-4** ([$AuCl_2(phen)]^+$), based on the motif Au(N^N). The zinc finger targets evaluated are also represented: the HIV-1 nucleocapsid protein NCp7 (ZnF2 boxed), along with the human transcription factor Sp1

Gln residue. Although the starting compound suffered reduction, the N-coordinated Au(I) is unusual [8]. Au(III)(N^N) and Au(III)(N^N^N) compounds were shown to cause zinc displacement in ZnF proteins but often followed by Au(III) to Au(I) reduction, leading to the formation of gold(I) fingers, as previously demonstrated for compounds such as [Au(dien)(N-donor)] [9] and [Au(terpy)Cl] [10].

Given the unique stability of the Au(C^N) class of compounds, and the unusual interaction products with proteins obtained in the few cases reported in the literature, we explored the application of compound **II-1** as a Zn ejector targeting the HIV-1 nucleocapsid protein ("full" protein and F2, with Cys_3His domains) as well as the human transcription factor Sp1 (Cys_2His-2). The Au(III) compounds were also evaluated regarding the inhibition of the interaction between the full-length NCp7 ZnF and a model DNA sequence (SL2).

5.2 Experimental

5.2.1 *Materials*

$H[AuCl_4]$, was purchased from Acros. 2,2′-Bipyridine (bipy), 4,4′-dimethyl-2,2′-bipyridine (dmbipy) and 1,10-phenanthroline (phen) monohydrate were obtained from Sigma-Aldrich. Deuterated solvents are from Cambridge isotopes or Sigma-Aldrich.

5.2.2 *Synthesis*

[Au(bnpy)Cl₂] (II-1). An adaptation of a published method was used [11]. In summary, $H[AuCl_4]\cdot 3H_2O$ (0.25 mmol, 98.5 mg) was dissolved in 5 mL of distilled water. To this solution, the ligand 2-benzylpyridine (0.25 mmol, 40.1 μL) was added. A yellow solid precipitated immediately. The suspension was left under reflux for 18 h and the solid turned white. The solid was isolated by filtration, washed with cold water and dried in vacuo. Anal. Calcd. for $C_{12}H_{10}AuCl_2N$ (436.09 g mol^{-1}): C 33.05%, H 2.31%, N 3.21%. Found: C 33.16%, H 1.87%, N 3.47%. ESI(+)-QTOF-MS ($[M\text{-}Cl + CH_3CN]^+$, m/z): 441.0446, calculated 441.0438. ^{1}H NMR (dmso-d6, 400 MHz): δ 4.50 (dd, 2H), 7.08 (td, 1H), 7.20 (td, 1H), 7.26 (dd, 1H), 7.42 (dd, 1H), 7.72 (ddd, 1H), 8.00 (dd, 1H), 8.27 (td, 1H), 9.19 (dd, 1H). Crystals suitable for single crystal X-ray analysis were obtained by recrystallization from acetonitrile.

[AuCl₂(bipy)][PF₆] (II-2). An adaptation of a published method was used [12]. In summary, $K[AuCl_4]$ (0.25 mmol, 95.0 mg) was dissolved in 3 mL of H_2O/CH_3CN (1:5). The ligand bipy, (0.25 mmol, 39.5 mg) was dissolved in 0.5 mL of CH_3CN and added to the gold(III)-containing solution. NH_4PF_6 (0.75 mmol, 124.6 mg) was added as a solid to the solution. The mixture was refluxed for 16 h. The solids were isolated by filtration and washed with cold water and dried in vacuo. Anal. Calcd. for $C_{10}H_8AuCl_2F_6N_2P$ (569.02 g mol^{-1}): C 21.11%, H 1.42%, N 4.92%. Found: C 21.08%, H 0.99%, N 5.00%. ESI(+)-QTOF-MS (M^+, m/z): 422.9755, calculated 422.9730. Crystals suitable for single crystal X-ray analysis were obtained by recrystallization from acetonitrile. ^{1}H NMR (dmso-d6, 400 MHz): δ 8.15 (ddd, 2H), 8.74 (dd, 2H), 8.94 (dd, 2H), 9.42 (dd, 2H).

[AuCl₂(dmbipy)][PF₆] (II-3) was synthesized following the same procedure used for compound **II-2**, using dmbipy instead of bipy. Anal. Calcd. for $C_{12}H_{12}AuCl_2F_6N_2P$ (597.07 g mol^{-1}): C 24.14%, H 2.03%, N 4.69%. Found: C 24.47%, H 1.49%, N 4.89%. ESI(+)-QTOF-MS (M^+, m/z): 453.0089, calculated 453.0015. ^{1}H NMR (dmso-d6, 400 MHz): δ 2.68 (s, 6H), 7.96 (dd, 2H), 8.78 (d, 2H), 9.21 (d, 2H).

[AuCl₂(phen)][PF₆] (II-4) was synthesized following the same procedure used for compound **II-2**, using 1,10-phenanthroline instead of 2,2′-bipy. Anal. Calcd. for $C_{12}H_8AuCl_2F_6N_2P$ (593.04 g mol^{-1}): C 24.30%, H 1.36%, N 4.72%. Found: C

24.08%, H 0.70%, N 4.73%. ESI(+)-QTOF-MS (M^+, m/z): 446.9707, calculated 446.9730. ^{1}H NMR (dmso-d6, 500 MHz): δ 8.26 (dd, 2H), 8.40 (s, 2H) 9.11 (dd, 2H), 9.33 (dd, 2H). Crystals suitable for single crystal X-ray analysis were obtained by recrystallization from acetonitrile.

Elemental Analysis data were acquired on an Elemental Analyzer CHNS-O 2400 Perkin Elmer. Mass spectra were acquired in a XEVO QTOF–MS instrument (Waters). The sample was dissolved in the smallest possible volume of DMSO and diluted in a 1:1 (v/v) mixture of water and acetonitrile containing 0.1% formic acid. NMR spectra (^{1}H, HSQC and HMBC) were acquired on a Bruker Avance III 400/500 MHz instrument.

NCp7 (F2 and "full") and Sp1 preparation. An adequate amount of apo-NCp7 was dissolved in water and zinc acetate was added (1.2:1 mol/mol per zinc binding domain). The pH was adjusted to 7.2–7.4 using a solution of NH_4OH. The final solution was incubated for 2 h at 37 °C prior to any experiments. The ZnF formation was confirmed by circular dichroism and mass spectrometry. The overall CD profile and species distribution in MS were in agreement with data previously reported for all ZnFs studied here.

5.2.3 X-Ray Crystal Structures

Single crystals of compounds **II-1**, **II-2** and **II-4** were analyzed on a Bruker ApexII CCD diffractometer using graphite monochromated Mo K_α ($\lambda = 0.71703$ Å) radiation. Least squares refinement was used to determine unit cell parameters (dimension and orientation matrices) obtained by θ-χ scans. The ApexII Suite ([APEX2 v2014.1-1 (Bruker AXS), SAINT V8.34A (Bruker AXS Inc., 2013)] was used for data integration and scaling.

The structures were solved using direct methods available on *ShelXT* and refined using a full-matrix least-squares technique on F^2 using *ShelXL* [13], as part of Olex2 [14]. Non-hydrogen atoms were refined anisotropically while hydrogen atoms were added in idealized positions and further refined based on the riding model. Figures were generated using Mercury 3.6.

5.2.4 Cyclic Voltammetry

Cyclic voltammograms were obtained in an Autolab EcoChemie PGSTAT20 potentiostat using tetrabutylammonium hexafluorophosphate (0.10 mol L^{-1} in dry dmf) as supporting electrolyte for all measurements. An Ag/AgCl electrode was used as pseudo-reference, a Pt wire was used as auxiliary electrode and a glassy carbon electrode was used as the working electrode. Stock solutions containing 5 mmol L^{-1} of compounds **II-1**, **II-2**, $H[AuCl_4]\cdot 3H_2O$ and the ligands bnpy and 2,2′-bipy were used for measurements. Every solution evaluated was prepared in dry, degassed dmf containing 0.10 mol L^{-1} tetrabutylammonium hexafluorophosphate. The scan rates evaluated were 10, 25, 50, 100 and 200 mV s^{-1}.

5.2.5 Interaction with Model Biomolecules

5.2.5.1 *N*-Acetyl-*L*-Cysteine

NMR. To a 10 mmol L^{-1} solution of compound **II-1** in dmso-d6, *N*-Ac-Cys was added (1:1 mol/mol). ^{1}H NMR spectra were acquired on a Bruker Avance III 400/500 MHz immediately after mixing and over time (one spectrum was acquired every 15 min for 4 h). Spectra were also acquired after 24 and 96 h. *MS*. To a 10 mmol L^{-1} of compound **II-1** in acetonitrile was added 1 eq of *N*-Ac-Cys dissolved in acetonitrile. MS was acquired immediately after mixing.

5.2.5.2 GSH

NMR. To a 25 mmol L^{-1} solution of compound **II-1** in dmso-d6, GSH was added (1:1 mol/mol). ^{1}H NMR spectra were acquired on a Varian 300 MHz immediately after mixing and over time (one spectrum was acquired every 15 min for 4 h; a final spectrum was acquired after 24 h).

5.2.5.3 Gly-L-His

NMR. To a 10 mmol L^{-1} solution of compound **II-1** in dmso-d6, Gly-L-His was added (1:1 mol/mol). ^{1}H NMR spectra were acquired on a Bruker Avance III 400/500 MHz immediately after mixing and over time (one spectrum was acquired every 15 min for up to 4 h; a final spectrum was acquired after 24 h).

5.2.6 Targeting NCp7 F2 and the Full-Length NCp7 ZnF by MS

For MS experiments, 1 mmol L^{-1} reaction mixtures (1:1 mol/mol of Au(III) complex per ZnF core) were prepared in water/acetonitrile mixtures at pH 7.0 (adjusted using NH_4OH). The reaction solutions were incubated at room temperature for 0, 2, 6 and 24 h. The samples were sprayed using a final concentration of ~100 μmol L^{-1}. Experiments were carried out on an Orbitrap Velos from Thermo Electron Corporation operated in positive mode. Samples (25 μL) were diluted with methanol

(225 μL) and directly infused at a flow rate of 0.7 μL/min using a source voltage of 2.30 kV. The source temperature was maintained at 230 °C throughout.

5.2.7 *NCp7 (Full)/SL2 (DNA) Interaction and Inhibition*

NCp7 and SL2 binding control experiment. A range of concentrations of NCp7 (aa 1-55) were mixed with 100 nmol L^{-1} 3′-fluorescein-labeled hairpin SL2 DNA (sequence GGGGCGACTGGTGAGTACGCCCC) in a final volume of 50 μL buffer containing 1.25 mmol L^{-1} NaCl, 0.125 mmol L^{-1} HEPES pH 7.2 in a 96-well black, low-binding microplate (Greiner). Fluorescence Polarization (FP) readings were recorded immediately on a CLARIOstar microplate reader.

Inhibition of NCp7-SL2 binding experiment. Serially diluted samples of compounds **II-1**, **II-2** and **II-4** (stocks prepared in dmf) were incubated with 5 μmol L^{-1} NCp7 for 30 min before addition of SL2 DNA. The final concentration of DNA was 100 nmol L^{-1} in a total volume of 50 μL. The solution was buffered with 0.125 mmol L^{-1} HEPES at pH 7.2 containing 1.25 mmol L^{-1} NaCl. Experiments were carried out in a 96-well black, low-binding microplate (Greiner). The NCp7 concentration was chosen from the NCp7-SL2 binding experiment such that 90% of SL2 was bound. FP readings were recorded immediately after the addition of SL2.

5.3 Results and Discussion

5.3.1 *Single-Crystal X-Ray Diffraction*

Crystals suitable for single-crystal X-ray diffraction were obtained for compounds **II-1**, **II-2** and **II-4**. The structures of compounds **II-2** (Figs. 5.2 and 5.9) and **II-4** (Figs. 5.3 and 5.10) are new and will be further discussed. The structures of compounds **II-1** (Fig. 5.11, also previously deposited under CCDC entry HOSHEE) and **II-3** (published elsewhere [12], CCDC entry KUMYEX) will be brought in for comparison. Bond distances of the first coordination spheres of compounds **II-1** to **II-4** are summarized in Table 5.1. A full description of the crystallographic data and refinement parameters obtained for compounds **II-1**, **II-2** and **II-4** is given in Table 5.2.

The longest Au-N bond in the series (2.046(7) Å) was observed for [Au(bnpy)Cl_2], while the shortest (2.020(10) Å) was part of [$AuCl_2$(dmbipy)]$^+$. The high pK_a of dmbipy (caused by the inductive effect of the methyl groups) can explain the shorter bond distances. In terms of Au-Cl bonds, the longest observed in the series appears for the chloride *trans* to the Au-C bond (2.348(3) Å), as expected by the strong *trans* effect caused by the carbanion. Even the Au-Cl *trans* to the pyridine ring in compound **II-1** (2.324(2) Å) is still longer than the average Au-Cl found on the

Table 5.1 Selected structural parameters for the organometallic compound II-1 in comparison to the Au(N^N) compounds II-2 to II-4

Compound	Distance/Å				Angles/deg	
Au(C^N)	Au-N	Au-C	Au-Cl	Au-Cl′	N-Au-C	Cl-Au-Cl
II-1	2.046(7)	2.060(7)	2.324(2)	2.348(3)	86.9(3)	91.73(9)
Au(N^N)	Au-N	Au-N′	Au-Cl	Au-Cl′	N-Au-N	Cl-Au-Cl
II-2	2.040(2)	2.036(2)	2.271(1)	2.269(1)	80.79(8)	88.65(3)
II-3	2.027(9)	2.020(10)	2.257(3)	2.250(4)	80.7(4)	88.9(1)
II-4	2.036(2)	2.032(2)	2.251(1)	2.255(1)	81.75(9)	89.28(3)

Table 5.2 Crystallographic data and refinement parameters for compounds **II-1**, **II-2** and **II-4**

Parameters	$[Au(bnpy)Cl_2]$	$[AuCl_2(bipy)]^+$	$[AuCl_2(phen)]^+$
Formula	$C_{12}H_{10}AuCl_2N$	$C_{10}H_8AuCl_2N_2PF_6$	$C_{10}H_8AuCl_2N_2PF_6$
M_r	436.08	569.02	593.04
T/K	150(2)	150(2)	150(2)
Crystal system	Monoclinic	Monoclinic	Orthorhombic
Space group	$P2_1/n$	$P2_1/n$	*Pbca*
a/Å	8.0827(12)	6.738(3)	12.9983(7)
b/Å	8.4639(12)	14.809(6)	15.2709(10)
c/Å	17.392(3)	14.728(6)	15.5153(10)
β (°)	91.420(3)	100.605(10)	90
V/Å^3	1189.4(3)	1444.5(11)	3079.7(3)
Z	4	4	8
ρ_{calc}/g cm^{-3}	2.435	2.616	2.558
μ/mm^{-1}	12.786	10.726	10.068
F_{000}	808	1056	2208
Radiation	Mo Kα (λ = 0.71703 Å)		
2θ range/°	4.69–54.79	3.93–56.56	4.88–56.56
Reflections collected	9594	28,635	15,573
Independent reflections	2675	3566	3822
R_{int}, R_σ	0.0377, 0.0375	0.0329, 0.0186	0.0268, 0.0231
Parameters, restraints	145, 0	199, 0	217, 0
S (on F^2)	1.161	1.020	1.007
R_1,wR_2 [$I \geq 2\sigma(I)$]	0.0382, 0.0749	0.0154, 0.0327	0.0190, 0.0381
R_1, wR_2 (all data)	0.0488, 0.0777	0.0186, 0.0336	0.0273, 0.0399
$\Delta\rho_{max}/\Delta\rho_{min}$/e·Å^{-3}	2.30/−2.33	1.31/−0.70	1.08/−0.56

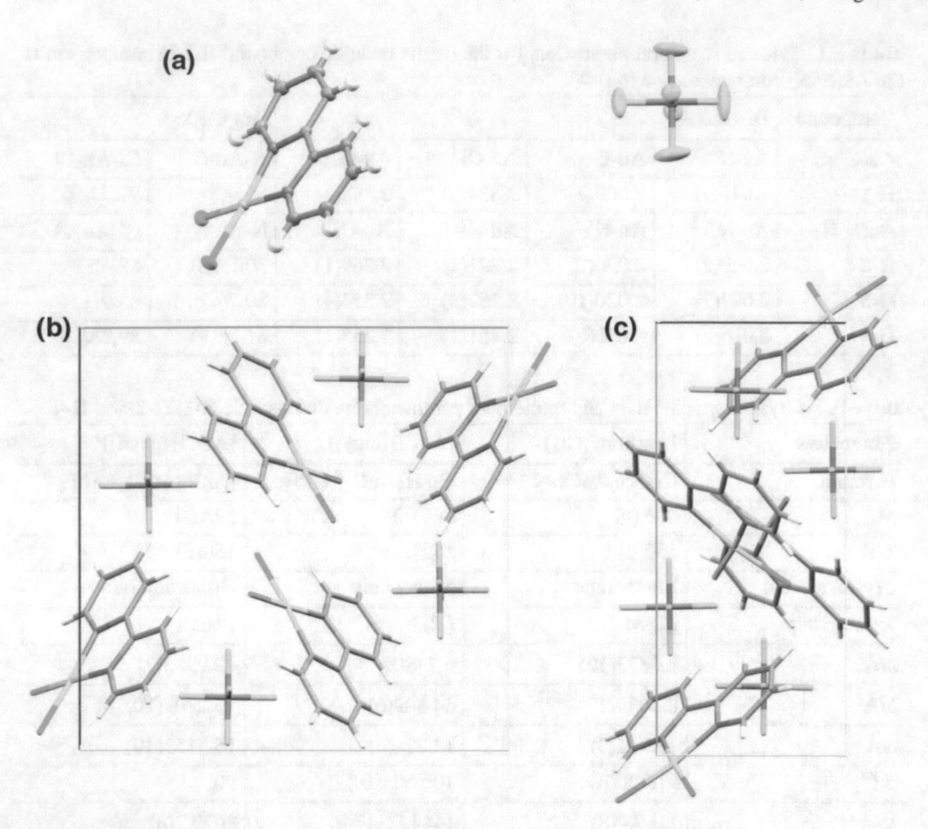

Fig. 5.2 **a** Ellipsoid view (50% probability) of the asymmetric unit of compound $[AuCl_2(bipy)]PF_6$. **b** Packing view along the a axis. **c** Packing view along the c axis

Au(N^N) series. In the Au(N^N) series, $[AuCl_2(bipy)]^+$ has slightly longer Au-Cl bonds than the other compounds (2.27 vs. 2.25 Å). As comparison, the compound $[AuCl_4]^-$, which has the chlorides perfectly positioned in the vertices of a square centered in the Au atom, has an average Au-Cl bond distance of 2.26–2.29 Å [15].

In terms of short contacts, $[AuCl_2(bipy)]^+$ has the most complex network, relying on interactions with the $[PF_6]^-$ counterion and also with other $[AuCl_2(bipy)]^+$ subunits to keep the crystal structure together (Fig. 5.9). Two $[PF_6]^-$ units appear positioned right above and below the coordination plane of $[AuCl_2(bipy)]^+$, which leads to the shortest anion-π interactions observed in the structure [Cg1···F3 (2.983 Å), Cg1···F1^i (3.019 Å), (i) = ½ – x, – ½ + y, ½ – z]. In addition, one of the pyridine rings appears surrounded by $[PF_6]^-$ units, with many C-H···F contacts adding up to the stability of the crystal [C3-H3···F2^{ii}, C4-H4···F5^{ii}, C4-H4···F4^{iii}, C5-H5···F3^{iv}, C6-H6···F5^{iv}, with the corresponding C3···F interactions ranging from 3.133 to 3.433 Å. Symmetry codes (ii) = −x, 1 − y, −z; (iii) = 1 − x, 1 − y, −z; (iv) = 1 + x, y, z]. Additionally, interactions between $[AuCl_2(bipy)]^+$ subunits take place through three Cl···H-C interactions, with corresponding Cl···C interactions in the range of 3.487

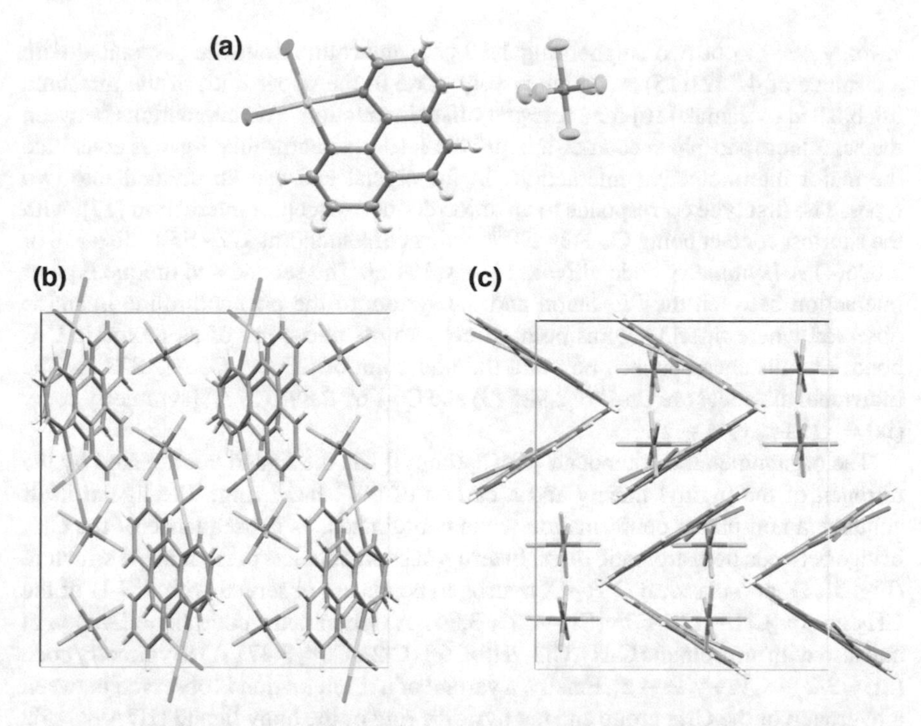

Fig. 5.3 **a** Ellipsoid view (50% probability) of the asymmetric unit of compound $[AuCl_2(phen)]PF_6$. **b** Intermolecular interactions, highlighting the hydrophobic CH···F and the anion-π interactions with the hexafluorophosphate counterion. **c** Packing view along the c axis

to 3.530 Å [Cl1···H4′-C4′^{v}, Cl1···H3′-C3′^{v}, Cl2···H5′-C5′^{vi}. Symmetry codes (v) = ½ − x, ½ + y, ½ − z; (vi) = −x, 1 − y, 1 −z].

The structure of compound **II-3**, reported elsewhere [12], differs dramatically from compounds **II-2** and **II-4** in terms of interaction between the $[AuCl_2(N\hat{}N)]^+$ group with the PF_6 counter-ion. One of the faces of $[PF_6]^-$ is positioned right above the $[AuCl_2(dmbipy)]^+$ unit. As consequence, F···Au (shortest: 3.11Å) and F···N (shortest: 2.97 Å) interactions are also present, in addition to the F···C (shortest: 2.99 Å) and F···HC (shortest: 2.52 Å). Regarding the chlorides, interactions with both aliphatic (Cl···H_3C, shortest: 2.9 Å) and aromatic (Cl···HC, shortest: 2.75 Å) hydrogens can be observed.

In the structure of $[AuCl_2(phen)]^+$ (Fig. 5.3), Au(III) deviates from planarity (as determined based on the four coordinating atoms) by 0.018 Å (r.m.s). Despite the highly symmetrical nature of the hexafluoridophosphate counter-ion, this unit does not show any disorder. The molecular packing of the $[AuCl_2(phen)]PF_6$ crystal is shown in Fig. 5.10. Despite the square-planar coordination environment around Au(III) and the presence of the highly conjugated and planar 1,10-phenanthroline ligand, π interactions have little relevance to the stabilization of the crystal. The shortest π-like interaction between the centroids [Cg2···Cg3^{vii}; symmetry code: (vii)

½+x, y, ½ – z] of two neighboring 1,10-phenanthroline rings are associated with a distance of 4.2521(15) Å, which is very close to the upper limit of the threshold established by Janiak [16] for a relevant offset interaction. The interactions between the hexafluorophosphate counter-ion and the 1,10-phenanthroline ligands constitute the major intermolecular interactions in the crystal and can be divided into two types. The first type corresponds to an anion donor π-acceptor interaction [17], with the shortest contact being C2-H2···F5^{viii} with a corresponding C2···F5^{viii} distance of 3.096(4) Å [symmetry code viii = x, ½ – y, ½+z]. The second and unique type of interaction between the PF_6 anion and the system of the phenanthroline ligand is observed where fluorine atoms point directly to the mid-point of an aromatic C-C bond. The distance between F6^{ix} and the mid-point of C5 and C6 is 2.822 Å. The individual distances are C5-F6^{ix} 2.925 (3) and C6-F6^{ix} 2.894 (3) A° [symmetry code: (ix) = – ½+x, y, ½ – z].

The organometallic compound [$AuCl_2$(bnpy)] has the ligand coordinated by the nitrogen of the pyridyl moiety and a carbon of the benzyl ring. The ligand itself acquires a non-planar conformation when coordinated, as consequence of the CH_2 bridge between both aromatic rings. In terms of short contacts present in the structure (Fig. 5.11), non-classical C-H···Cl hydrogen bonds are observed with a C-H of the CH_2 group [Cl1···H7B-C7^{x} (Cl1···C7^{x}, 3.841 Å) [Symmetry code (x) = 1+x, y, z] and also with an aromatic C-H [(Cl2···H6-C6^{xi} (Cl2···C6^{xi}, 3.471 Å) [Symmetry code (xi) = 3/2 – x, ½+y, ½ – z]. Finally, a variant of a π interaction is observed between a hydrogen of the CH_2 group and the pyridine ring of the bnpy ligand [H7A···Cg5^{xii} (C7···Cg5^{xii} 3.398 Å). Symmetry code (xii) = ½ – x, ½+y, ½ – z].

5.3.2 NMR

The full NMR characterization of compound [Au(bnpy)Cl_2] was not reported in the literature so far. For that reason, compound **II-1** was fully characterized by solution NMR techniques (Fig. 5.12). The most distinct spectroscopic feature of [Au(bnpy)Cl_2] is the non-equivalency of the hydrogens of the CH_2 group when bnpy acts as a bidentate ligand. This feature allows us to probe coordination and, indirectly, the oxidation state of Au (as bnpy can be a bidentate ligand for Au(III) only).

The labile chlorides found in compounds **II-1** to **II-4** are susceptible to solvent replacement as previously described in the literature, for example, for [$AuCl_2$(phen)]$^+$ in water [18]. How quickly this reaction happens depends on the strength of the Au-Cl bond and also on the *trans* effect induced by the bidentate ligand. Figures 5.13, 5.14, 5.15 and 5.16 show the ^{1}H NMR spectra of compounds **II-1** to **II-4** in dmso-d6 immediately after dissolution and after 24–72 h. Considering the lower lability of the bidentate C^N/N^N ligands over the chloride, three species are expected to coexist in dmso: [$AuCl_2L$]$^+$, [AuClL(dmso)]$^{2+}$ and [AuL$(dmso)_2$]$^{3+}$. All three species were identified for compound **II-2**, while [AuLCl_2]$^+$ and a ligand replacement product were identified for compound **II-3**. Of all the Au(N^N) complexes, [$AuCl_2$(phen)]$^+$ (**II-4**) has the largest amount of the chloride replacement product shortly after dissolution (most likely $[AuL(dmso)_2]^{3+}$), which becomes the only species in solution after 72 h. The cyclometallated compound **II-1**, on the other hand, shows no evidence of extensive chloride replacement by dmso-d6, even 48 h after dissolution.

Compounds **II-1** to **II-4** were also compared in terms of ^{15}N chemical shifts. This data was obtained indirectly, using the $\{^1H, ^{15}N\}$ HMBC technique and samples with natural abundance of ^{15}N (Fig. 5.17). The ^{15}N signal observed for $[Au(bnpy)Cl_2]$ (227.54 ppm) does not differ significantly from that of $[AuCl_2(bipy)]^+$ (230.30 ppm) and $[AuCl_2(dmbipy)]^+$ (224.26 ppm). The signal obtained for $[AuCl_2(phen)]^+$ (244.32 ppm) is the one that differs the most from the rest of the series, as consequence of the different electronic system (delocalized over three fused aromatic rings) of phen.

5.3.3 *Redox Stability by Cyclic Voltammetry (CV)*

The redox stability of organometallic compound **II-1** was further studied in comparison to compounds **II-2** and the precursor $H[AuCl_4]$ (Fig. 5.4). Reduction of compound **II-2** is observed at −0.42 V and for $H[AuCl_4]$ at −0.45. On the other hand, for compound **II-1**, the reduction process happens at −1.21 V, in agreement to previously published data [19]. The −0.8 V difference indicates an increased stability of compound **II-1** towards reduction, a very desirable property for a compound that will be exposed to the Cys-rich biological media. Comparisons with the free ligand and the solvent (dmf) are shown in Fig. 5.18.

5.3.4 *Targeting Model Biomolecules*

5.3.4.1 *N*-Acetyl-*L*-Cysteine (N-Ac-Cys)

N-Ac-Cys was selected because it is a biologically relevant model S-donor. The interaction of $[Au(bnpy)Cl_2]$ (**II-1**) with *N*-Ac-Cys was followed using NMR (Fig. 5.19) and mass spectrometry (Fig. 5.20). By ^1H NMR, it was possible the determine whether bnpy is still coordinated as bidentate ligand to Au(III). The 6-membered metallacycle can still be observed even after 24 h of incubation with *N*-Ac-Cys, which can be followed based on the coupling of the hydrogens on the CH_2 group. The two hydrogens are non-equivalent only when bnpy acts as a bidentate ligand. The MS spectrum acquired for the interaction between compound **II-1** and *N*-Ac-Cys shows very interesting features that are representative of the overall reactivity of this compound with Cys-containing biomolecules. The most abundant peak (527.06 m/z) corresponds to the simple ligand-replacement product, [Au(bnpy)(*N*-Ac-Cys)]$^+$. More interestingly, it is also possible to observe the C-transferred product bnpy-*N*-Ac-Cys with high abundance at 331.11 m/z. This unique product is evidence of an Au-mediated C-transfer to the S-donor (Cys arylation).

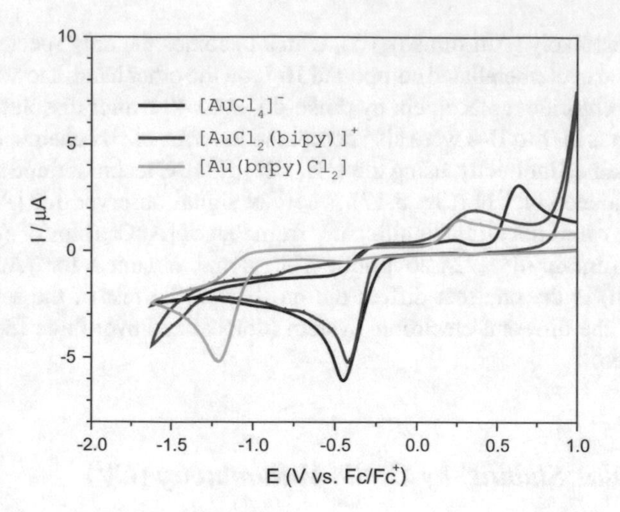

Fig. 5.4 Cyclic voltammogram (vs. Fc/Fc$^+$) of the organometallic compound **II-1** in comparison to the Au(N^N) compound **II-2** and the gold precursor $[AuCl_4]^-$. Tetrabutylammonium hexafluorophosphate (0.10 mol L^{-1} in dry dmf) was used as supporting electrolyte in all measurements. An Ag/AgCl electrode was used as pseudo-reference, a Pt wire was used as auxiliary electrode and a glassy carbon electrode was used as the working electrode

5.3.4.2 Glutathione (GSH)

As the most abundant intracellular S-containing peptide, GSH represents a relevant biological target. Using NMR, we probed the redox stability of compound **II-1** in presence of a high concentration of GSH (25 mmol L^{-1}), much higher than the intracellular total concentration of 7 mmol L^{-1} [20]. Figure 5.21 shows the spectra acquired for the interaction of $[Au(bnpy)Cl_2]$ in presence of GSH (t = 0 and up to 24 h after mixing). The metallacycle is still stable even after 24 h as observed by the presence of the non-equivalent hydrogens of the CH_2 group (in the 4.5 ppm region). The stability of the [Au(bnpy)] moiety in presence of GSH observed by ^{1}H NMR agrees with the CV electrochemical data of the free compound. Figure 5.22 shows the MS spectrum obtained for the interaction between compound **II-1** $[Au(bnpy)Cl_2]$ and GSH. The signal at 671.11 m/z corresponds to the major Au-containing product observed, $[Au(bnpy)(GSH)]^+$, which indicates chloride replacement by GSH. Furthermore, oxidized GSH ($GSSH^+$) was also observed at 613.15 m/z, which may correlate to Au(III) reduction to some extent. As a minor product, the species $[Au(bnpy)]_2$-GSH^+ was also identified at 1034.48 m/z.

5.3.4.3 Gly-L-His

Besides Cys residues, ZnF proteins also have His residues coordinated to Zn. Gly-L-His was selected as a model to His binding. The compound **II-1** was mixed with

Gly-L-His and the reaction was followed by ^{1}H NMR (Fig. 5.23). Compound **II-1** is less reactive with Gly-L-His than with the Cys-containing GSH and *N*-Ac-Cys. No changes are observed in the hydrogen signals of [Au(bnpy)Cl_2] even 24 h after mixing. Some changes were observed over time for the hydrogens H2 and H5 of the imidazole ring [21] of histidine, indicating an interaction.

5.3.5 Targeting Zinc Fingers

5.3.5.1 NCp7 (ZnF2)

The interaction of compounds **II-1** to **II-4** with NCp7 (ZnF2) was followed over time using MS (Figs. 5.24, 5.25, 5.26 and 5.27). The typical reactivity behavior of Au(III) complexes was observed for compounds **II-2** to **II-4** when interacting with NCp7 (ZnF2) at t = 0 (immediately after interaction). Reduction to different extensions was always observed, and the bidentate N-donor was not observed coordinated to Au. Compound **II-2** ([AuCl_2(bipy)]$^+$) interacts immediately with NCp7 F2 and the species oxiF (4+ at 555.99 and 5+ at 444.99 m/z) and Au(I)$_2$oxiF (3+ at 872.29, 4+ at 654.47 and 5+ at 523.77 m/z) represent the main observable products. Compound **II-3** [AuCl_2(dmbipy)]$^+$ also reacts fast, but AuF is the main gold-containing product observed here (2+ at 1209.97, 3+ at 806.98, 4+ at 605.48 and 5+ at 484.89 m/z). Oxidized finger is also observed immediately (oxiF^{2+} at 1110.98 and oxiF^{5+} at 445.00 m/z). Compound **II-4** ([AuCl_2(phen)]$^+$) was the least reactive compound among the Au(N^N) series since an aurated ZnF species, Au-ZnF (4+ at 621.21 m/z), was still observed at t = 0. AuF (4+ at 605.23 m/z) and oxiF (4+ at 556.24 m/z) were also identified. After 48 h, further auration was observed, with Au(I)$_2$F (3+ at 872.97, 4+ at 654.97 m/z) being identified, accompanied by AuF (4+ at 605.48, 3+ at 807.31 m/z).

On the other hand, compound **II-1** demonstrated a unique reactivity. At t = 0 (Fig. 5.24a), the chlorides were lost but bnpy remained coordinated to Au(III). Zinc displacement was also observed. Three related species were identified immediately after incubation: Au(III)(bnpy)-F (4+ at 647.50 m/z, 3+ at 876.15 m/z), Au(III)(bnpy)-ZnF (4+ at 663.49 m/z, 3+ at 884.31 m/z) and Au(III)(bnpy)-Zn$_2$F (4+ at 678.97 m/z, 3+ at 909.64 m/z). After 48 h further auration is observed, as observed by the presence of species such as ([Au(III)(bnpy)]$_3$-F)$^{4+}$ at 829.27 and ([Au(III)(bnpy)]$_3$-F)$^{3+}$ at 1105.36 m/z. Even more interestingly, the S-arylated product (first observed based on the interaction of compound **II-1** with *N*-Ac-Cys) was also identified here, after 24 h (Fig. 5.24b). The product (bnpy)-oxiapoF (5+ at 478.81 and 2+ at 1195.52 m/z) gives strong evidence for the Au(III)-mediated C-bond transfering to cysteine residues. Therefore, the following reactivity trend was established based on the reactivity of the Au(III) compounds with NCp7 (ZnF2):

$$[Au(bnpy)Cl_2] \ll [AuCl_2(phen)]^+ < [AuCl_2(dmbipy)]^+ < [AuCl_2(bipy)]^+$$

5.3.5.2 Targeting the Full-Length NCp7 Zinc Finger

Two compounds were selected for further studies using the full-length NCp7 ZnF as target. The organometallic [Au(bnpy)Cl_2] (Fig. 5.5a) was compared to [$AuCl_2$(bipy)]$^+$ (Fig. 5.5b).

Both compounds interact with the full-length NCp7 ZnF immediately after mixing. [$AuCl_2$(bnpy)] leads to the formation of oxiF (11+ at 584.64, 10+ at 642.90, 9+ at 714.33 and 8 + at 803.38). The chlorides are labilized and the cyclometallated adduct Au(bnpy)oxiF is observed (11+ at 617.73, 10+ at 679.50, 9+ at 754.78 and 8+ at 849.01 m/z). After 48 h (Fig. 5.28) another species appears, the Au-catalyzed C-S coupling product bnpy-oxiF (10+ at 659.92 and 9+ at 733.89 m/z). The mass spectrum of the species bnpy-oxiF^{10+}, in comparison to the theoretical isotopic distribution, is shown in Fig. 5.6.

[$AuCl_2$(bipy)]$^+$ also leads to the formation of oxiF (12+ at 536.00, with 11+, 10+ and 9+ charge states also being observed). On the other hand, AuF (11+ at 602.64, 10+ at 662.70 and 9+ 736.11 m/z) and Au_2F (11+ at 620.45, 10+ at 682.39 and 9+ at 757.99 m/z) are the main Au-containing species, with no N^N ligand coordination observable.

Combining the information acquired on the behavior of compound **II-1** ([Au(bnpy)Cl_2]) in the presence of the model S-donor *N*-Ac-Cys and also the ZnFs NCp7 (F2) and the full-length NCp7, the mechanism of C-S coupling shown in Scheme 5.1 can be proposed.

This is the first direct observation of a C-S coupling on a ZnF obtained by the interaction of a metal complex with this class of protein, representing a completely new mechanism of zinc displacement.

5.3.5.3 Targeting the Human Transcription Factor Sp1 (ZnF3)

Cleaner mass spectra were obtained for the interaction of compounds **II-1** to **II-4** with Sp1 ZnF3 (Fig. 5.7), as consequence of the higher intrinsic reactivity of the Cys_2His_2 motif (Sp1) when compared to the Cys_3His motif (NCp7). The Au(N^N) motif lead mainly to the formation of apoF (8+ at 421.72, 7+ at 481.82, 6+ at 561.96, 5+ at 674.15, 4+ at 842.44 m/z). [$AuCl_2$(phen)]$^+$, on the other hand, led to the appearance of AuF (7+ at 509.96, 6+ at 594.28 and 5+ at 712.94 m/z). Again, the most different behavior in the series was observed for [Au(bnpy)Cl_2], agreeing with the reactivity trend observed for NCp7 (ZnF2). The species Au(bnpy)-F (7+ at 533.97, 6+ at 622.80, 5+ at 747.16 and 4+ at 933.70 m/z) was the main product identified, as opposed to the Au(bnpy)-oxiF species identified when targeting NCp7 (ZnF2).

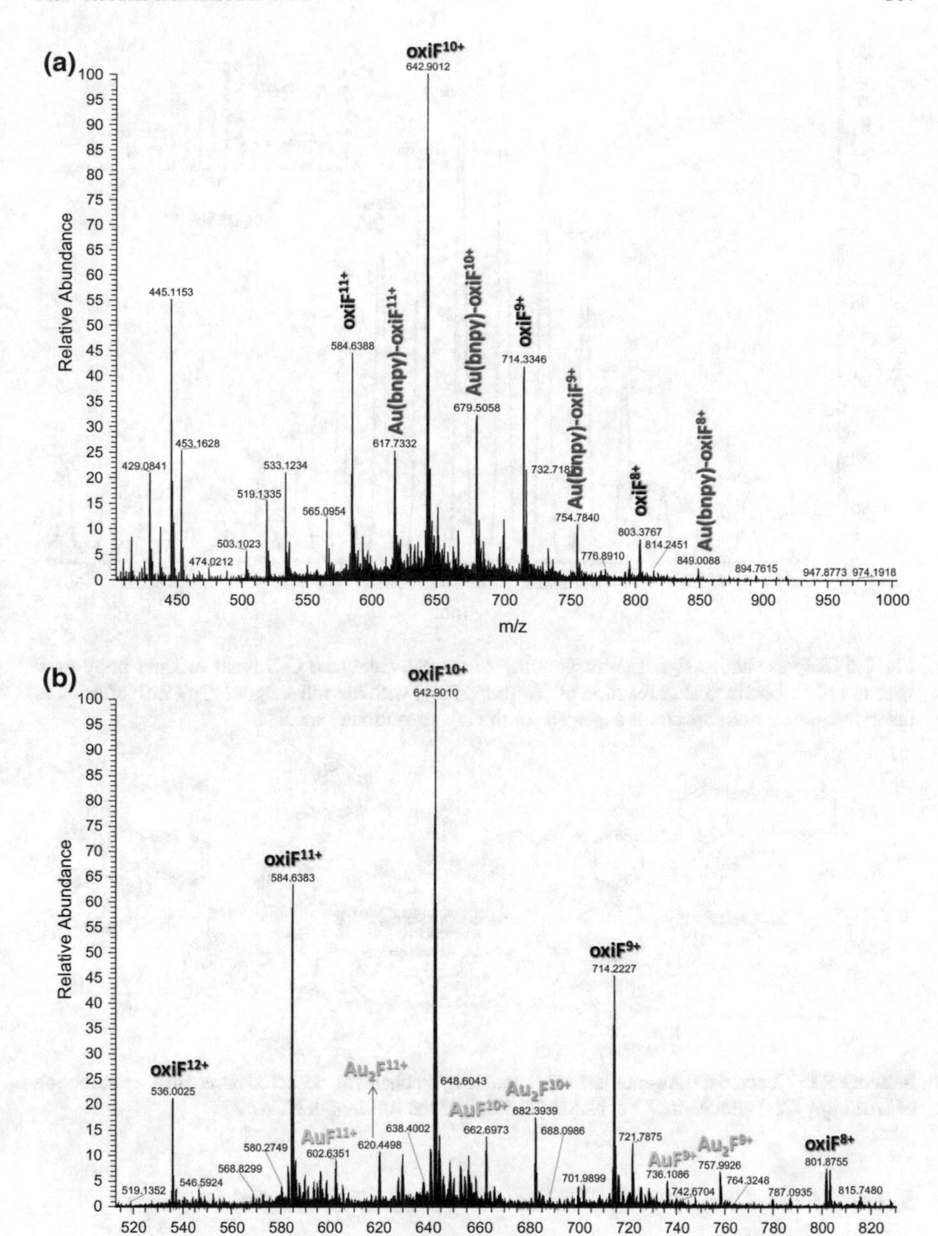

Fig. 5.5 Mass spectra obtained for the interaction of selected Au(III) compounds with the full-length NCp7 ZnF. **a** [Au(bnpy)Cl_2] in comparison to **b** [$AuCl_2$(bipy)]$^+$, immediately after mixing

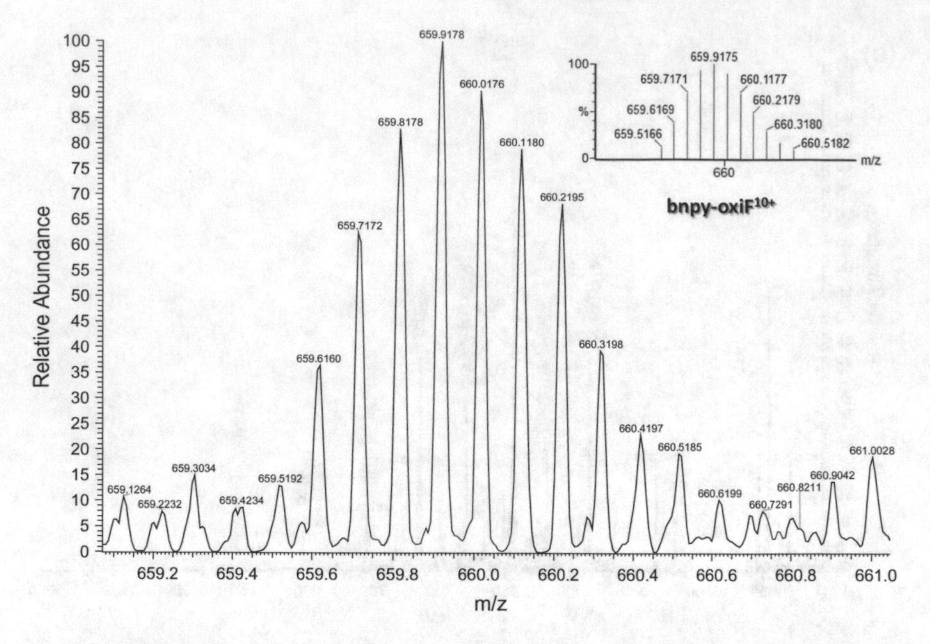

Fig. 5.6 Representative signal corresponding to the Au-catalyzed C-S bond product bnpy-oxiF species (10+), obtained after reaction of [Au(bnpy)Cl_2] with the full-length NCp7 ZnF after 24 h. Inset: theoretical mass spectrum expected for the aforementioned species

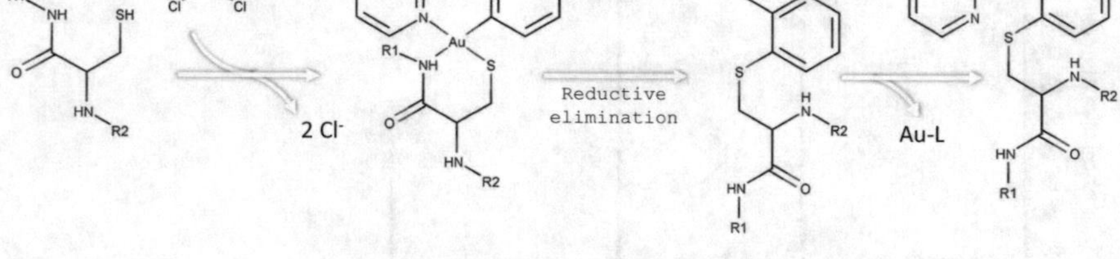

Scheme 5.1 Generalized Au-mediated C-S coupling mechanism, observed based on the interaction of [Au(bnpy)Cl_2] with *N*-Ac-Cys, NCp7 (ZnF2) and the full-length NCp7 ZnF

5.3.6 *Full-Length NCp7 Zinc Finger/SL2 DNA Interaction Inhibition*

Given the unique mechanism of Zn displacement observed for compound **II-1**, an experiment based on fluorescence polarization was set up to determine whether this compound is able to disrupt the interaction of NCp7 full-length zinc finger with the model DNA. The Au(N^N) compounds **II-2** to **II-4** were also studied in comparison (Fig. 5.8).

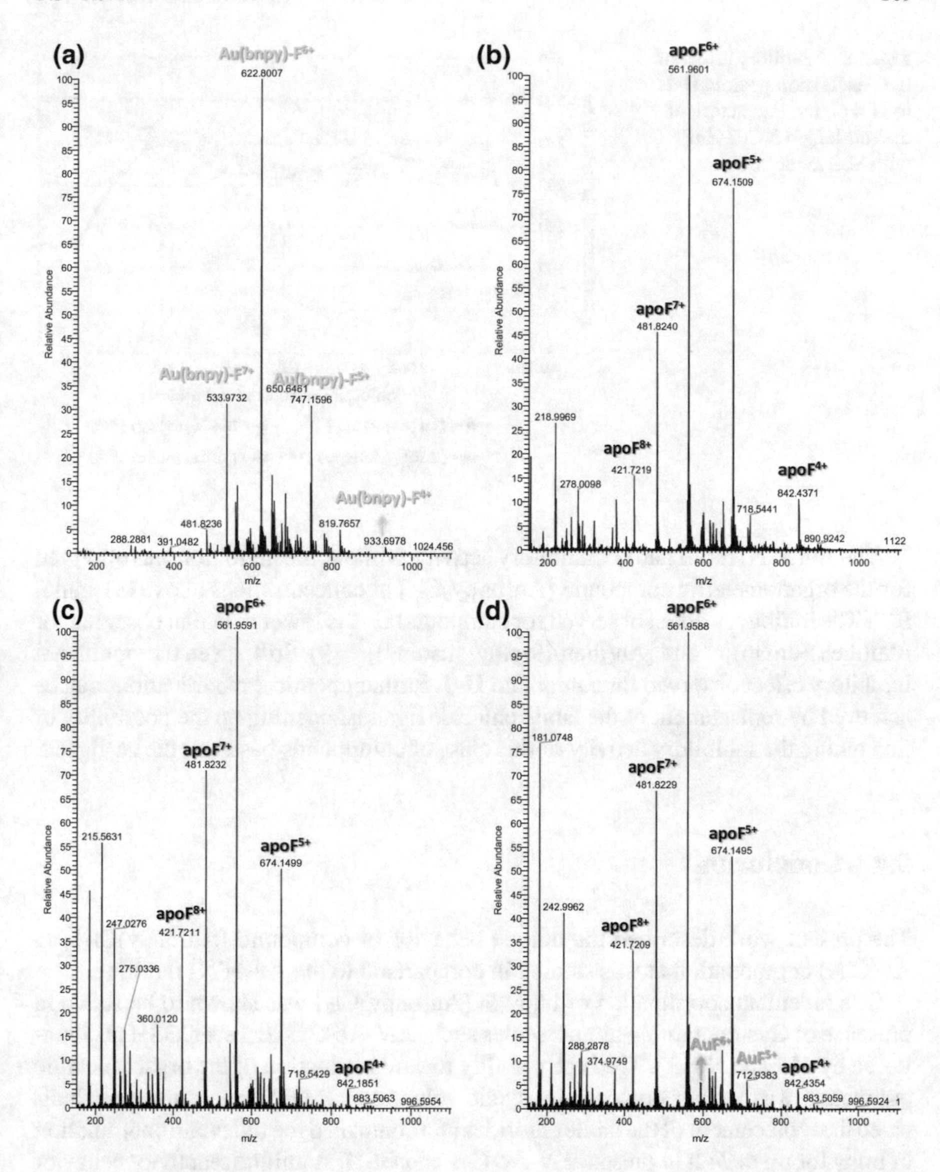

Fig. 5.7 Interaction of **a** $[Au(bnpy)Cl_2]$, **b** $[AuCl_2(bipy)]^+$, **c** $[AuCl_2(dmbipy)]^+$ and **d** $[AuCl_2(phen)]^+$ with the human transcription factor Sp1 (F3) immediately after mixing

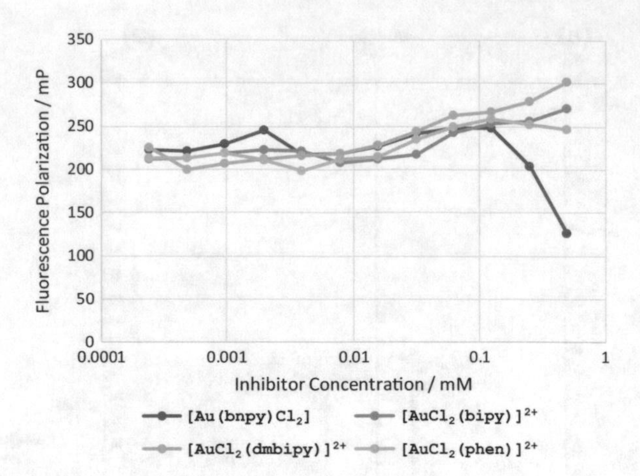

Fig. 5.8 Inhibitory effect of the Au(III) compounds **II-1** to **II-4** on the interaction of the full-length NCp7 ZnF with SL2 model DNA

The Au(N^N) motif had no inhibitory activity, while some inhibition was observed for the organometallic compound $[Au(bnpy)Cl_2]$ at concentrations above 0.1 mmol L^{-1}. The inhibitory effect observed for compound **II-1** is lower than that observed for $[Au(dien)(dmap)]^{3+}$ and $[Au(dien)(9\text{-ethylguanine})]^{3+}$ [9]. Still, given the significant inhibitory effect observed for compound **II-1**, further chemical modifications can be achieved by replacement of the labile chloride ligands, opening up the possibility of fine tuning the inhibitory activity of this class of compounds based on the co-ligand.

5.4 Conclusions

The present work described the unique behavior of compound $[Au(bnpy)Cl_2]$, an Au(C^N) compound, that was studied in comparison to the Au(N^N) motif.

The bidentate coordination of bnpy in $[Au(bnpy)Cl_2]$ was shown to be stable in presence of Cys-containing biomolecules such as *N*-Ac-Cys and even GSH (as monitored by MS and NMR). The high stability towards reduction of the organometallic compound was further observed by cyclic voltammetry. NMR measurements indicated the replacement of the labile chlorides and confirmed the bidentate coordination of bnpy for up to 24 h in presence *N*-Ac-Cys and GSH. A unique reactivity behavior was identified when $[Au(bnpy)Cl_2]$ was reacted with *N*-Ac-Cys and the reaction was followed by MS. The Au(III)-catalyzed C-S coupling product bnpy-*N*-Ac-Cys was formed. This unique reactivity was further explored targeting zinc finger proteins. For the HIV-1 protein NCp7 and the truncated model, zinc displacement was observed and the C-S coupling products were also identified.

On the other hand, the Au(N^N) compounds **II-2**, **II-3** and **II-4** were shown to react much faster with the target ZnFs evaluated here, undergoing reduction even at t = 0. Within the Au(N^N) series, $[AuCl_2(phen)]^+$ had lowest reactivity with ZnFs as demonstrated by MS experiments.

The unique mode of zinc displacement observed for the organometallic **II-1** also translated into a relevant inhibition of the full-length NCp7 zinc finger interaction with the SL2 model DNA, with no inhibitory effects observed for the Au(N^N) compounds.

Appendix

Crystal Structures

See Figs. 5.9, 5.10 and 5.11.

Nuclear Magnetic Resonance

See Figs. 5.12, 5.13, 5.14, 5.15, 5.16 and 5.17; Table 5.3.

Cyclic Voltammetry

See Figs. 5.18.

Interaction with Model Biomolecules

See Figs. 5.19, 5.20, 5.21, 5.22 and 5.23.

Table 5.3 ^{13}C assignments based on {^{1}H, ^{13}C} correlations identified by HSQC and HMBC NMR

δ ^{13}C	^{1}H correlation		Assignment
	HSQC	HMBC	
46.56	–	7.25 (dd), 7.99 (dd)	CH_2
124.98	7.71 (ddd)	9.17 (dd), 7.25 (dd)	4′
126.89	7.99 (dd)	4.48 (CH_2), 7.71 (dt) 8.26 (dt)	3′
127.42	7.07 (dt)	7.25 (dd)	4
128.45	7.19 (dt)	7.42 (dd)	5
129.11	7.25 (dd)	4.48 (CH_2), 7.07 (dt)	3
132.44	–	4.48 (CH_2), 7.42 (dd)	2′
133.20	7.42 (dd)	–	6
141.56	–	7.42 (dd), 7.25 (dd), 7.18 (dt), 7.07 (dt), 4.48 (CH_2)	2
143.76	8.27 (dt)	9.17 (dd)	5′
152.56	9.17 (dd)	8.27 (dt) 7.71 (dt)	6′
156.16	–	9.17 (dd), 8.27 (dt), 7.99 (dd), 4.48 (CH_2)	1′

Mass Spectrometry

Interaction with NCp7 (ZnF2)

See Figs. 5.24, 5.25, 5.26 and 5.27.

Interaction with the Full-Length NCp7 ZnF

See Fig. 5.28.

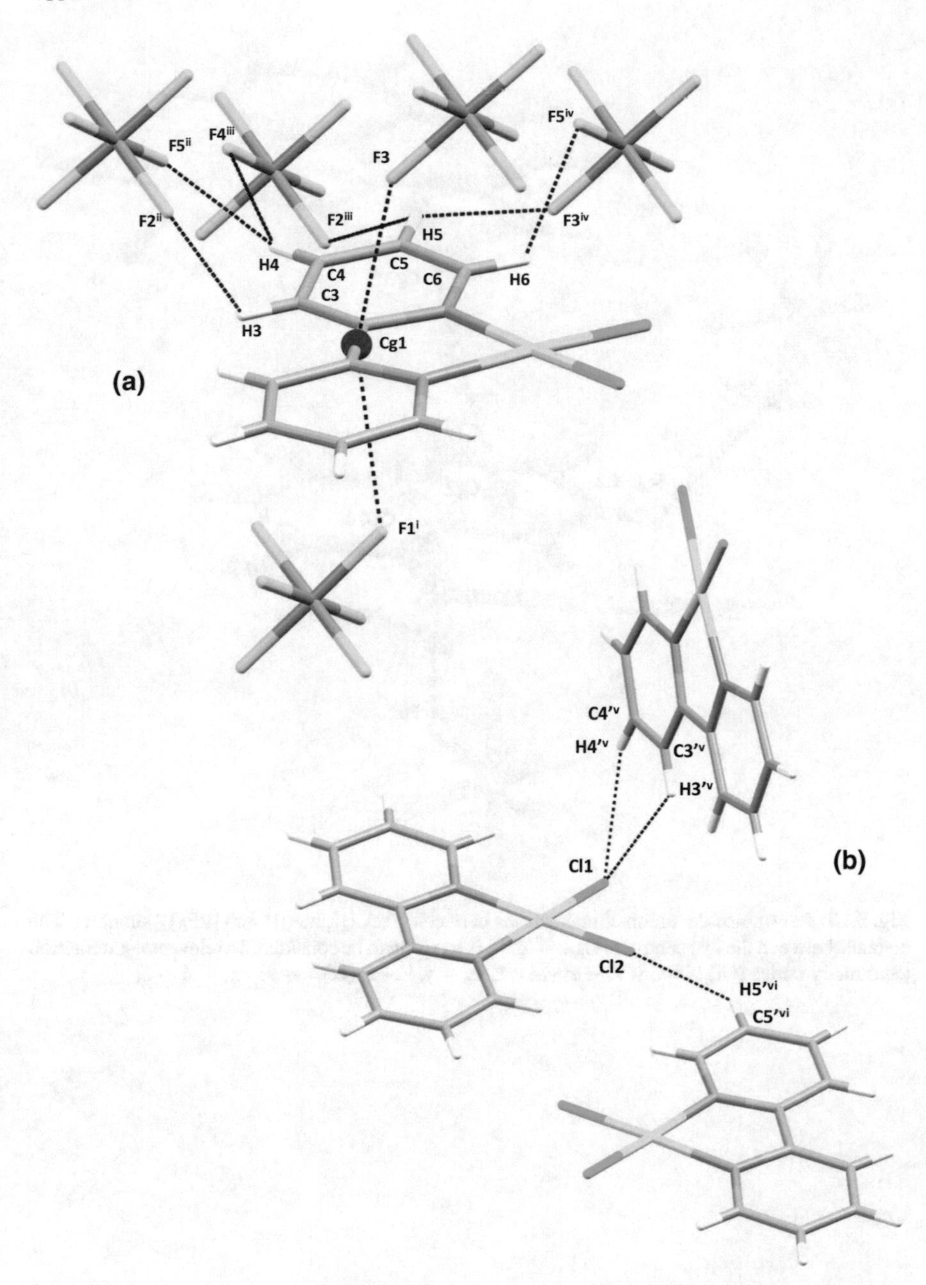

Fig. 5.9 Intermolecular contacts found in the structure of $[AuCl_2(bipy)][PF_6]$, emphasizing **a** contacts with the $[PF_6]^-$ counterion and **b** between $[AuCl(bipy)]^+$ subunits [Symmetry codes (i) = ½ − x, − ½ + y, ½ − z; (ii) = −x, 1 − y, − z; (iii) = 1 − x, 1 − y, −z; (iv) = 1 + x, y, z; (v) = ½ − x, ½ + y, ½ − z; (vi) = −x, 1 − y, 1 − z]

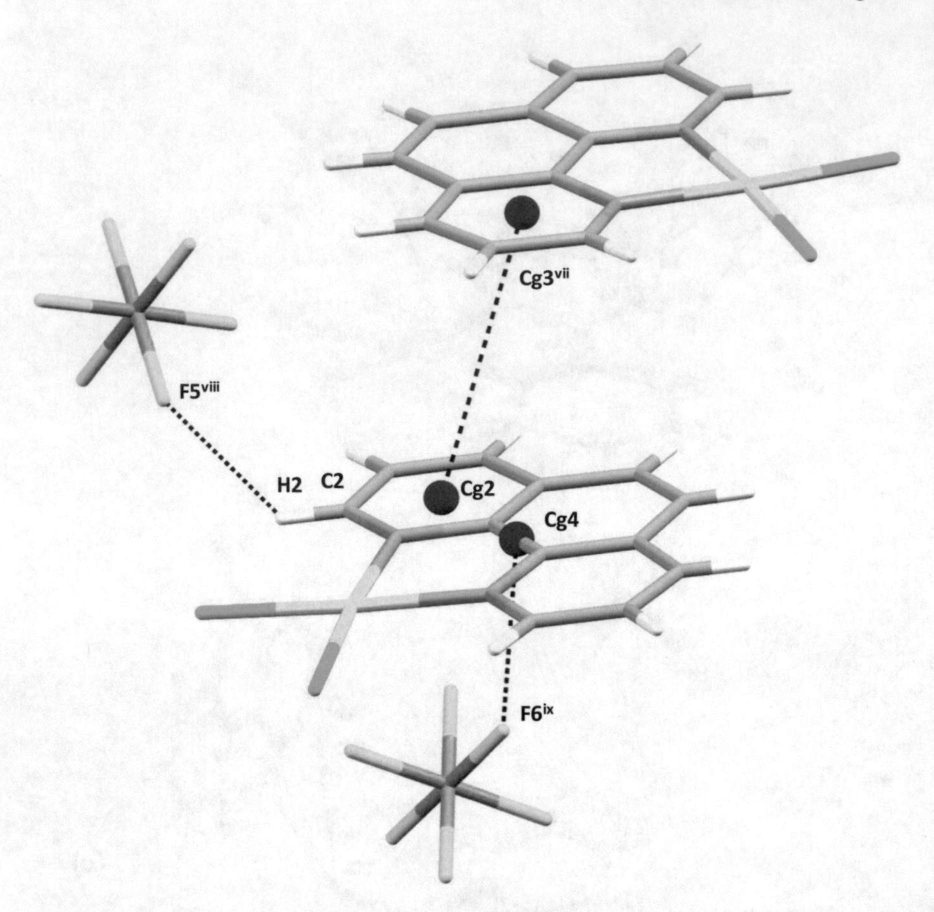

Fig. 5.10 Intermolecular anion-π interactions between $[AuCl_2(phen)]^+$ and $[PF_6]^-$ subunits. The distance between the two centroids Cg2 ···$Cg3^{vii}$ is too long to be considered a relevant π-interaction [Symmetry codes (vii) ½+x, y, ½ − z; viii = x, ½ − y, ½+z; (ix) = − ½+x, y, ½ − z]

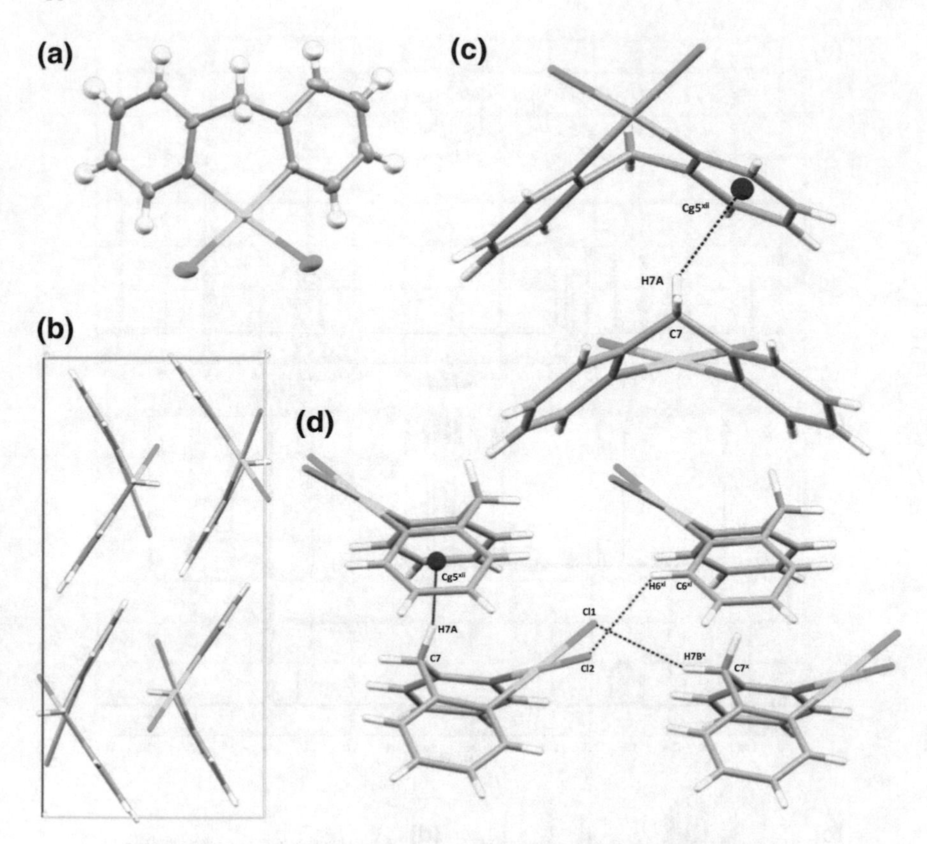

Fig. 5.11 **a** Ellipsoid view (50% probability) of the asymmetric unit of compound [Au(bnpy)Cl_2] **b** Packing view along the a axis. **c** Detailed view of the CH···π interaction. **d** Intermolecular interactions, highlighting the hydrophobic CH_2···Cl interaction between the CH_2 bridge and Cl ligands and also the CH···π interaction between the CH_2 bridge and the π-density on the pyridine ring of the bnpy ligand [Symmetry codes (x) = 1 + x, y, z; (xi) = 3/2 − x, ½ + y, ½ − z; (xii) = ½ − x, ½ + y, ½ − z]

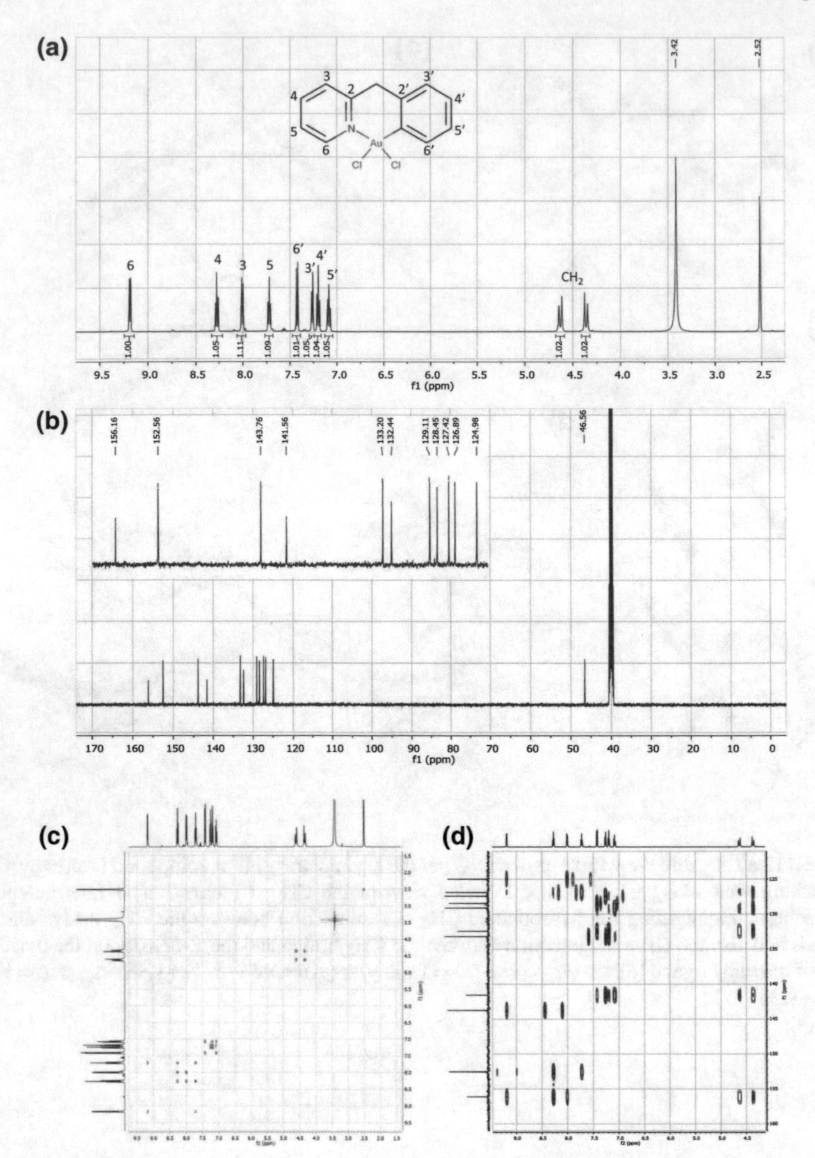

Fig. 5.12 **a** ^{1}H NMR, **b** ^{13}C NMR, **c** {^{1}H, ^{1}H}COSY and **d** {^{1}H, ^{13}C} HMBC of $[Au(bnpy)Cl_2]$

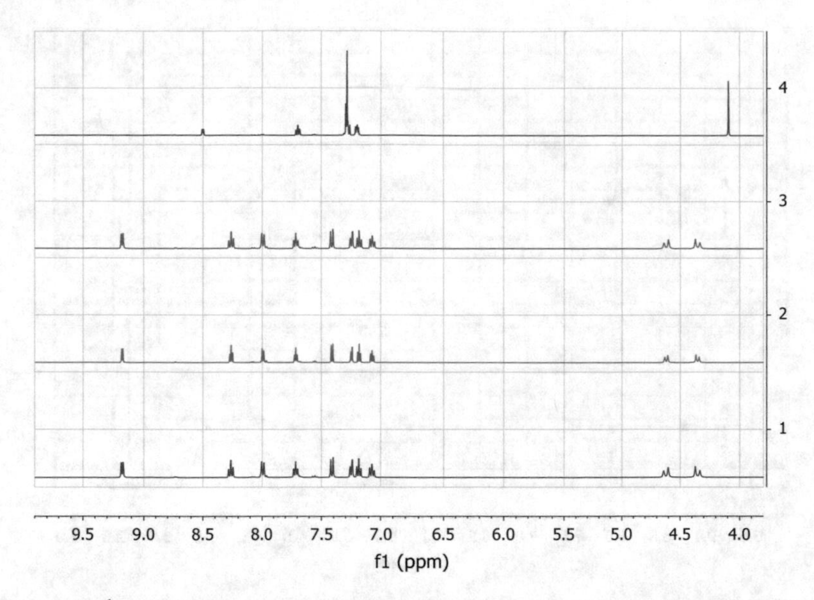

Fig. 5.13 (**1**) ^{1}H NMR of complex [Au(bnpy)Cl_2] freshly dissolved in dmso-d6 and spectra obtained (**2**) 24 h and (**3**) 48 h after dissolution. The spectrum of the free ligand bnpy is shown in (**4**) for comparison

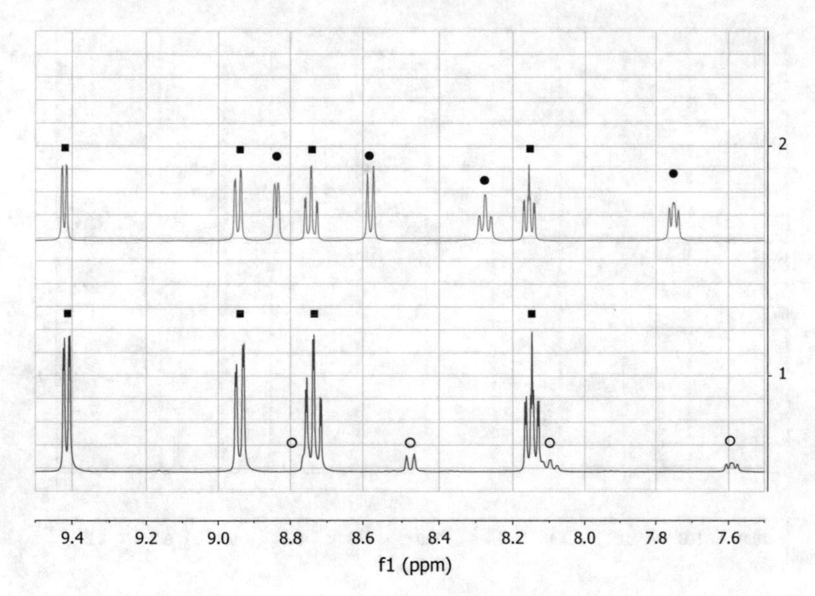

Fig. 5.14 ^{1}H NMR following the Cl replacement by dmso-d6 in the complex [$AuCl_2$(bipy)]PF_6. (**1**) Spectrum obtained for a freshly dissolved sample and (**2**) 24 h after dissolution. Three populations were identified, $[AuCl_2(bipy)]^+$ (■), $[AuCl(bipy)(dmso)]^{2+}$ (○) and $[Au(bipy)(dmso\text{-}d6)_2]^{3+}$ (●)

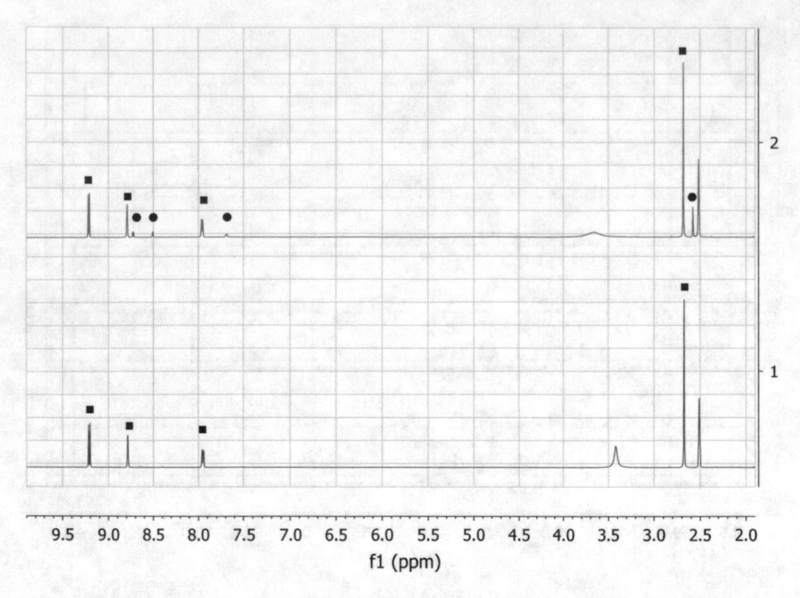

Fig. 5.15 ^{1}H NMR following the Cl replacement by dmso-d6 in the complex $[AuCl_2(dmbipy)]PF_6$. (**1**) Spectrum obtained for a freshly dissolved sample and (**2**) 24 h after dissolution. Two populations were identified, $[AuCl_2(dmbipy)]^+$ (■) and a chloride replacement product (●)

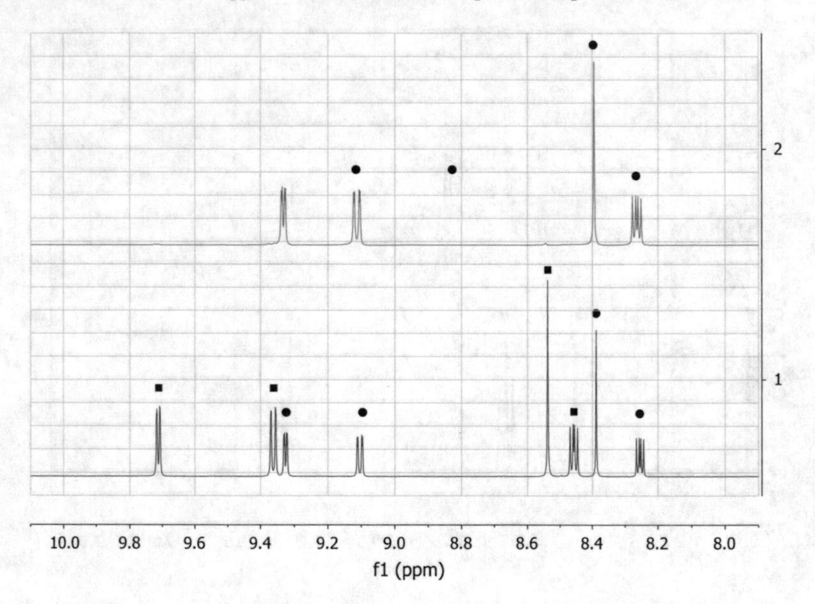

Fig. 5.16 ^{1}H NMR following the Cl replacement by dmso-d6 in the complex $[AuCl_2(phen)]PF_6$. (**1**) Spectrum obtained for a freshly dissolved sample and (**2**) 72 h after dissolution. Two populations were identified, $[AuCl_2(phen)]^+$ (■) and a chloride replacement product (most likely $[Au(phen)(dmso\text{-}d6)_2]^{3+}$, marked as ●)

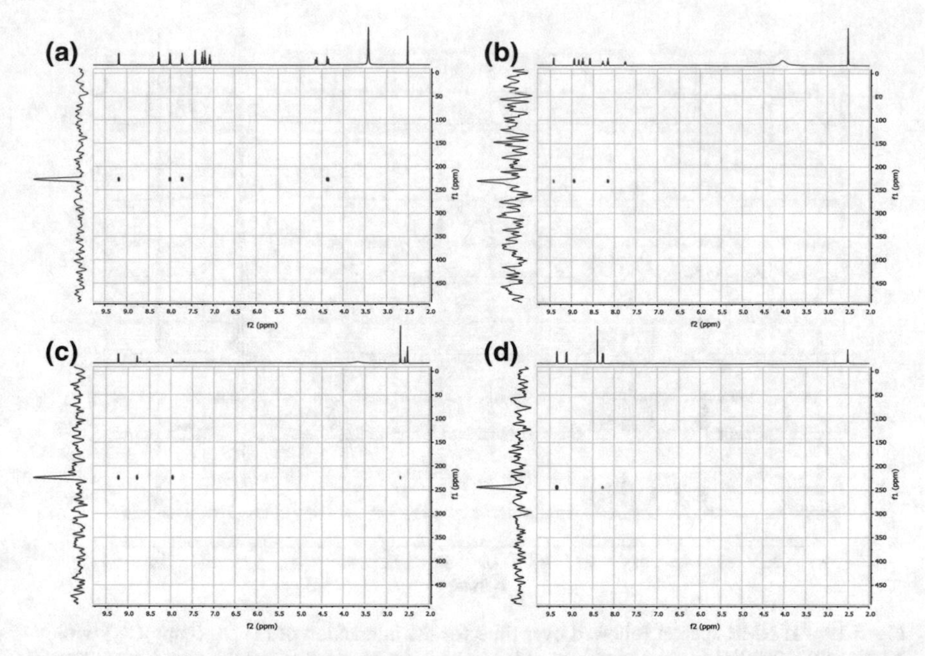

Fig. 5.17 {^{1}H, ^{15}N}HMBC spectrum of of compounds. **a** $[Au(bnpy)Cl_2]$, **b** $[AuCl_2(2,2'\text{-bipy})]^+$, **c** $[AuCl_2(dmbipy)]^+$ and **d** $[AuCl_2(phen)]^+$ obtained using samples with natural abundance of ^{15}N

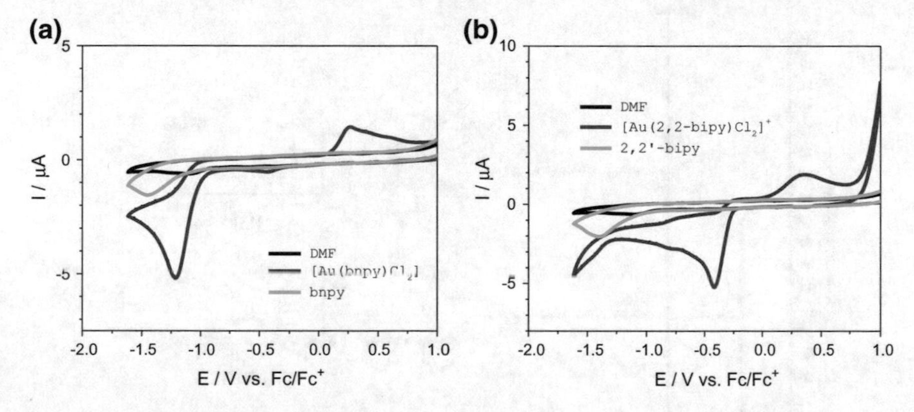

Fig. 5.18 **a** CV profile of the organometallic compound **II-1** in comparison to the free bnpy and **b** CV of$[AuCl_2(bipy)]^+$ in comparison to free bipy

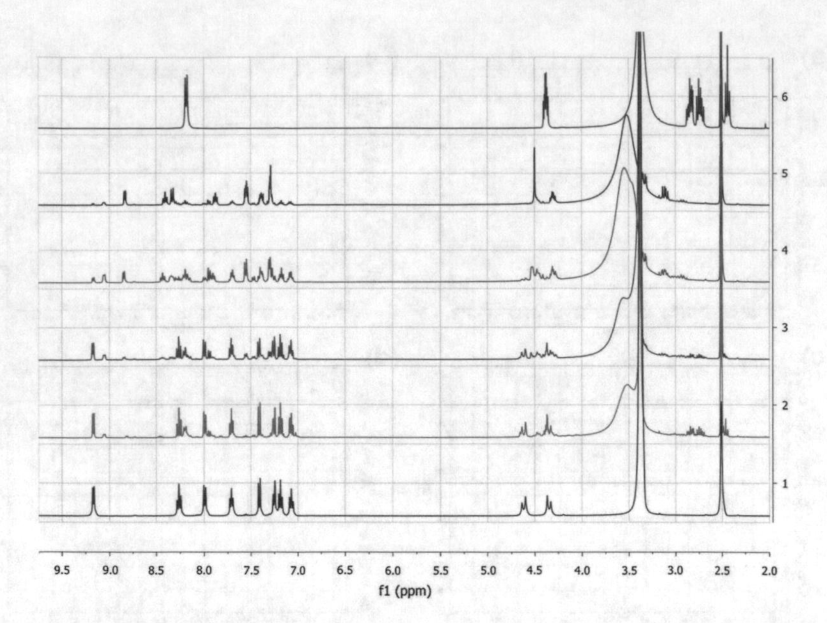

Fig. 5.19 ^{1}H NMR spectra followed over time for the interaction of (**1**) [Au(bnpy)Cl$_2$] with *N*-Ac-Cys (**2**) immediately after incubation, (**3**) after 4 h, (**4**) after 1 day and (**5**) after 4 days. The ^{1}H NMR spectrum of free *N*-Ac-Cys is given (**6**) for comparison

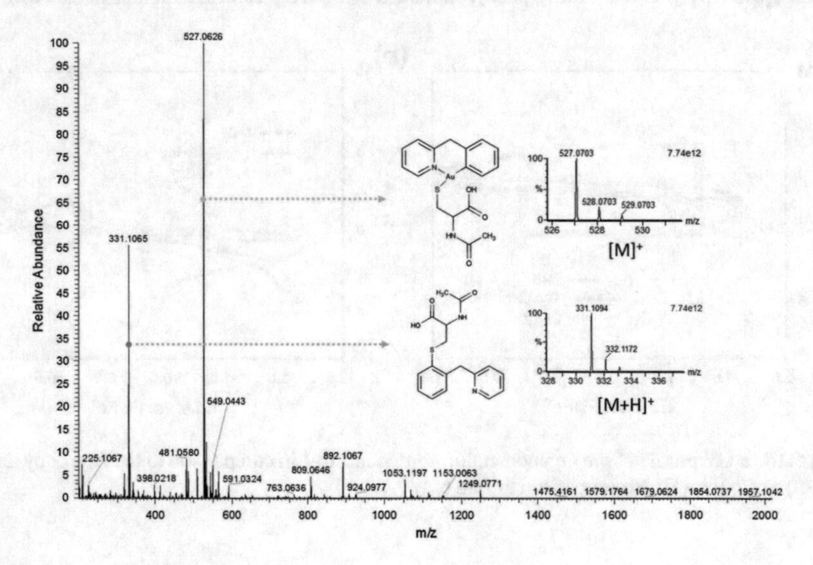

Fig. 5.20 MS spectrum obtained for the interaction between compound **II-1** [Au(bnpy)Cl$_2$] and *N*-Ac-Cys immediately after mixture

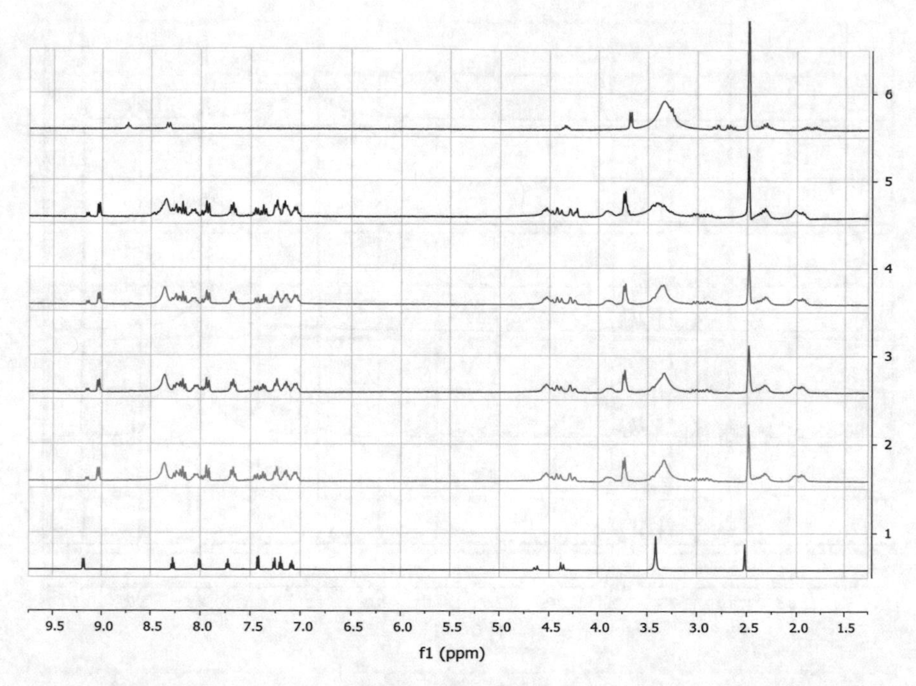

Fig. 5.21 ^{1}H NMR spectra followed over time for the interaction of (**1**) [Au(bnpy)Cl$_2$] with GSH (**2**) immediately after incubation, (**3**) after 1 h, (**4**) after 3 h and (**5**) after 24 h. The ^{1}H NMR spectrum of GSH is given (**5**) for comparison

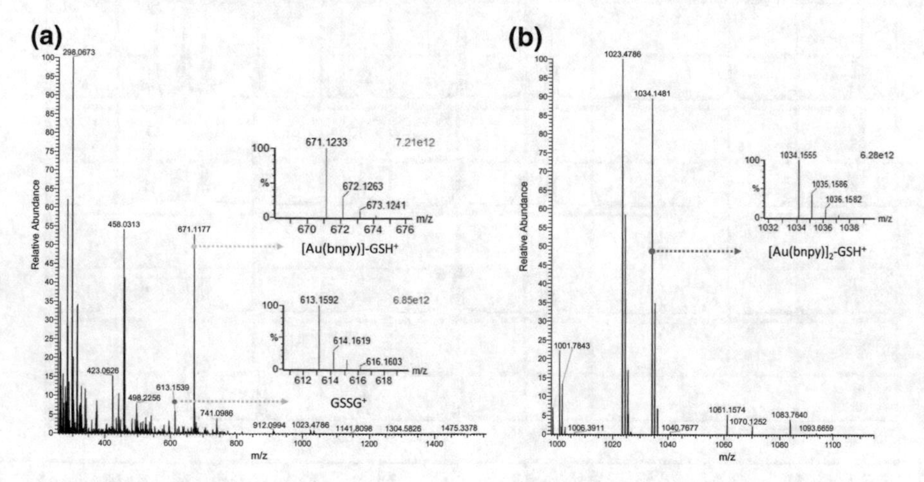

Fig. 5.22 MS spectrum obtained for the interaction between compound **II-1** [Au(bnpy)Cl$_2$] and GSH. **a** major and **b** minor species identified

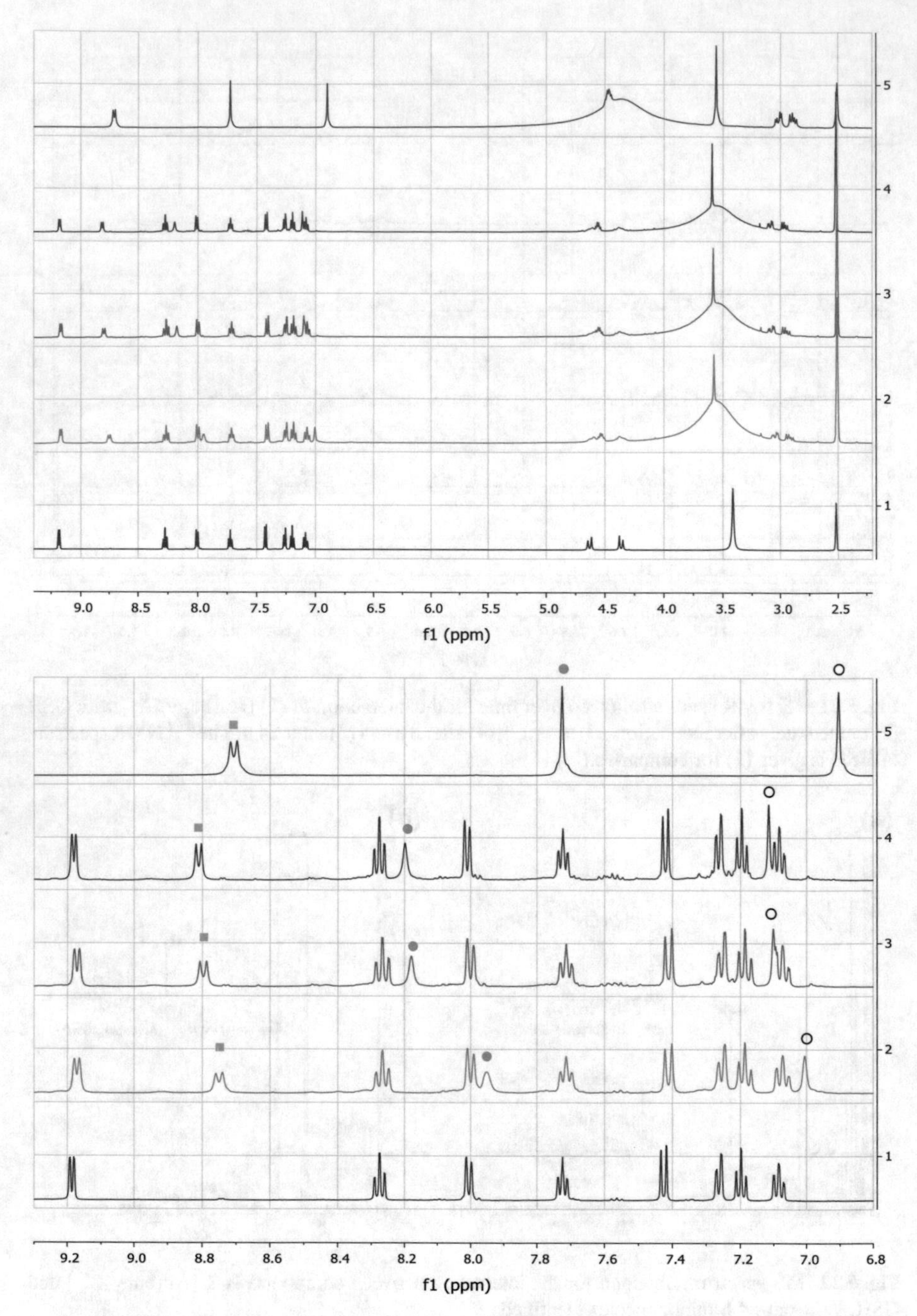

Fig. 5.23 ^{1}H NMR spectra followed over time for the interaction of (**1**) [Au(bnpy)Cl$_2$] with the dipeptide Gly-L-His (**2**) immediately after incubation, (**3**) after 3 h and (**4**) after 24 h. The ^{1}H NMR spectrum of Gly-L-His is given (**5**) for comparison. Top: Full spectrum; bottom: detailed view of the aromatic region. Hydrogens H2 and H5 of the imidazole ring found on Gly-L-His are marked in mustard and cyan respectively

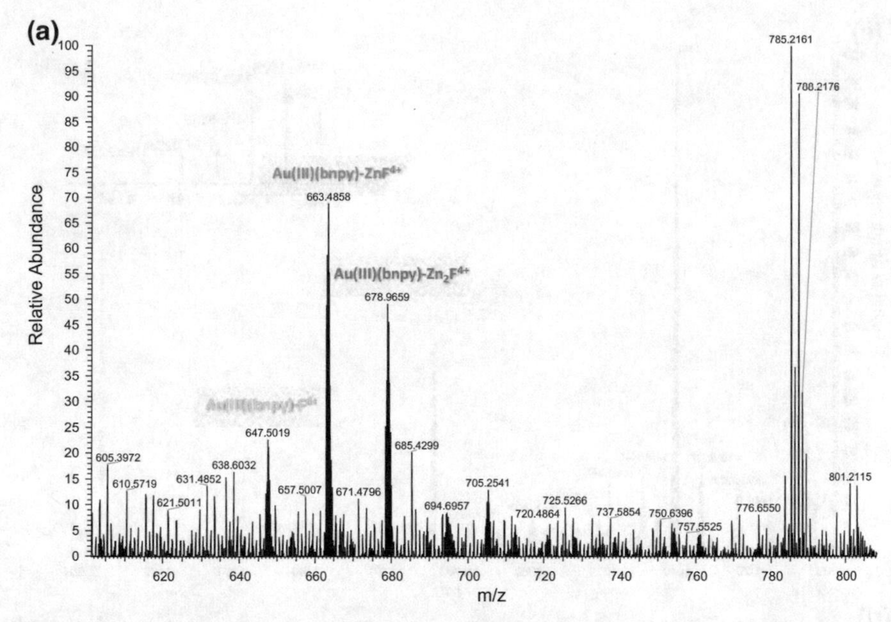

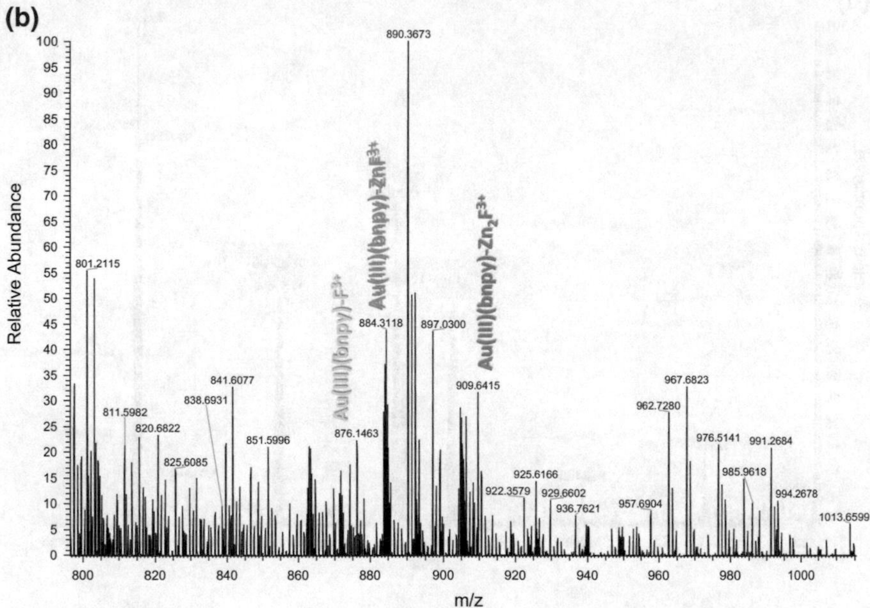

Fig. 5.24 Mass spectra obtained for the interaction [Au(bnpy)Cl_2] with NCp7 (ZnF2) **a** immediately after mixing and **b–d** 24 h after mixing, in different m/z regions

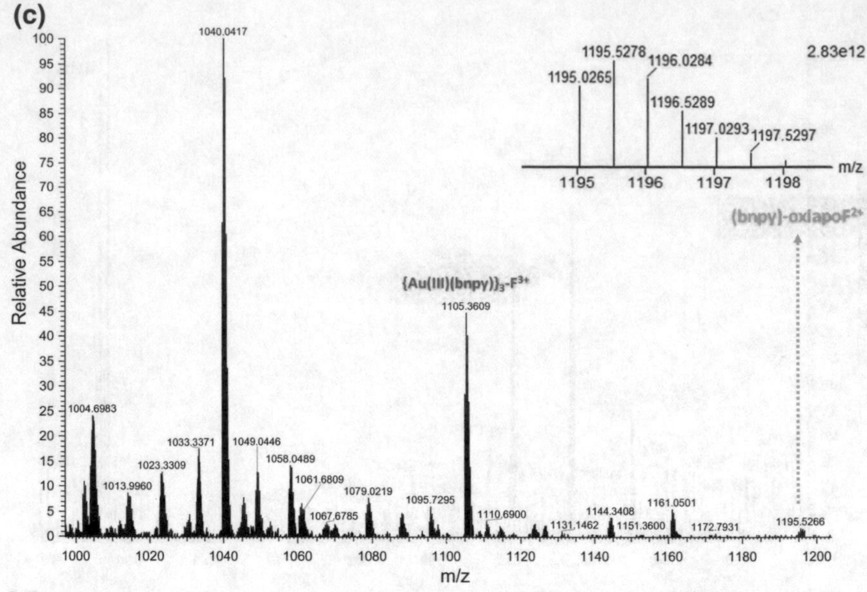

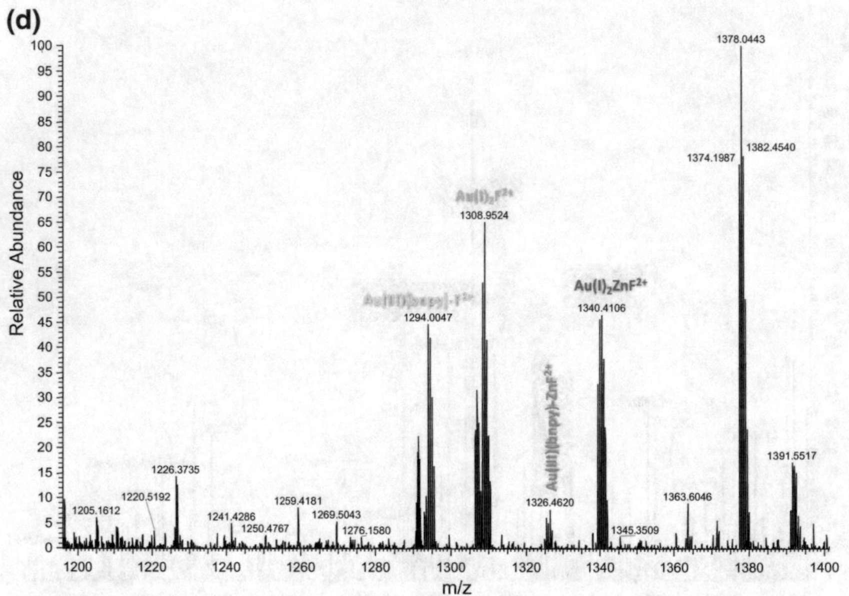

Fig. 5.24 (continued)

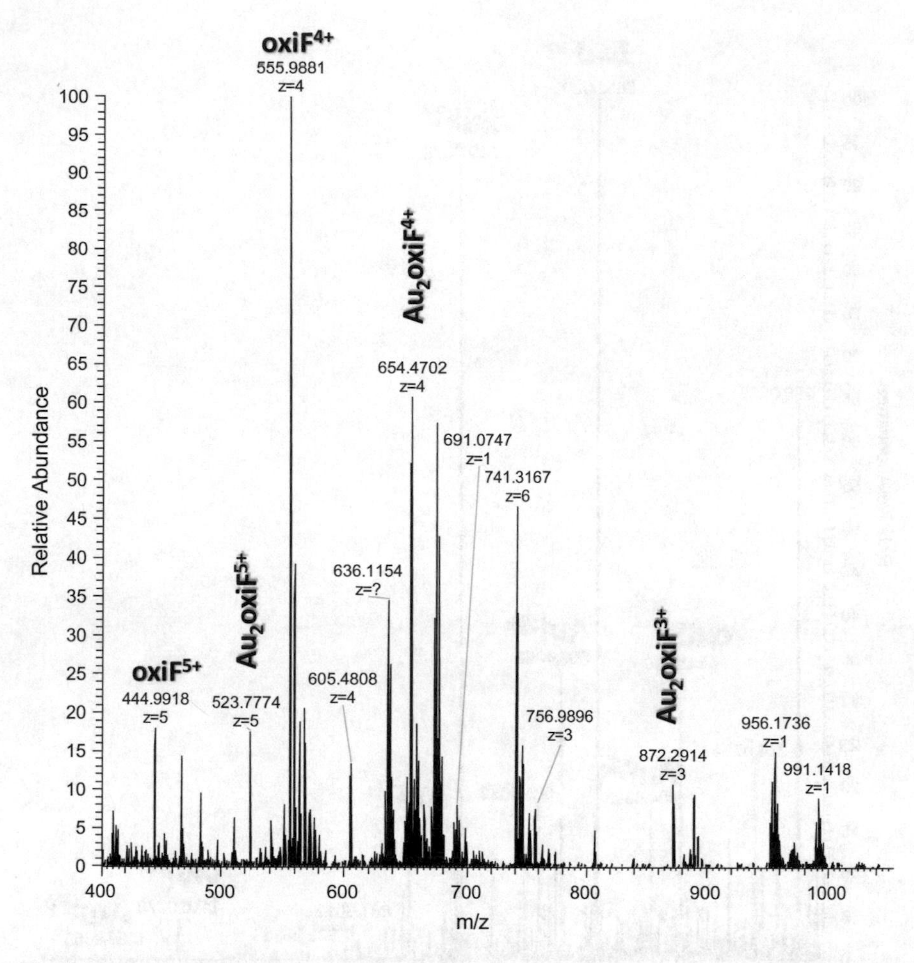

Fig. 5.25 Mass spectrum obtained for the interaction of $[AuCl_2(bipy)]^+$ with NCp7 (ZnF2) immediately after mixing

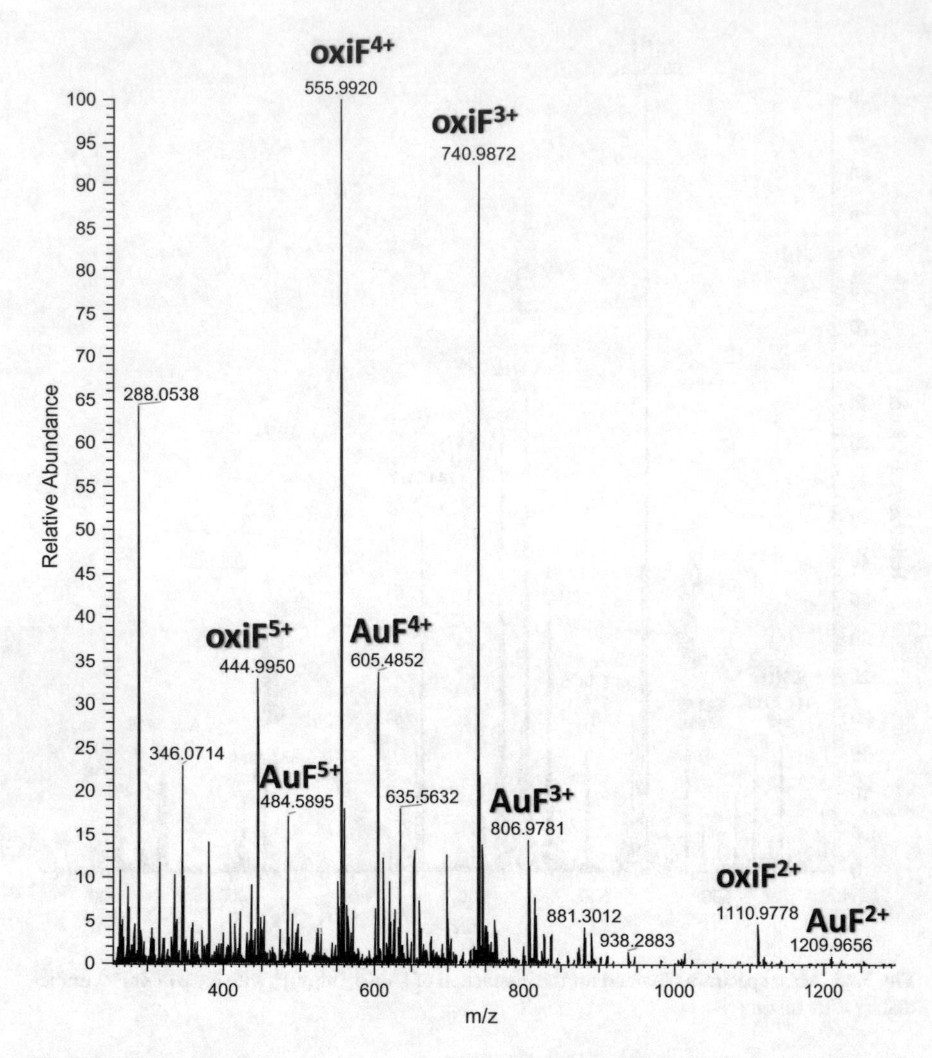

Fig. 5.26 Mass spectrum obtained for the interaction of $[AuCl_2(dmbipy)]^+$ with NCp7 (ZnF2) immediately after mixing

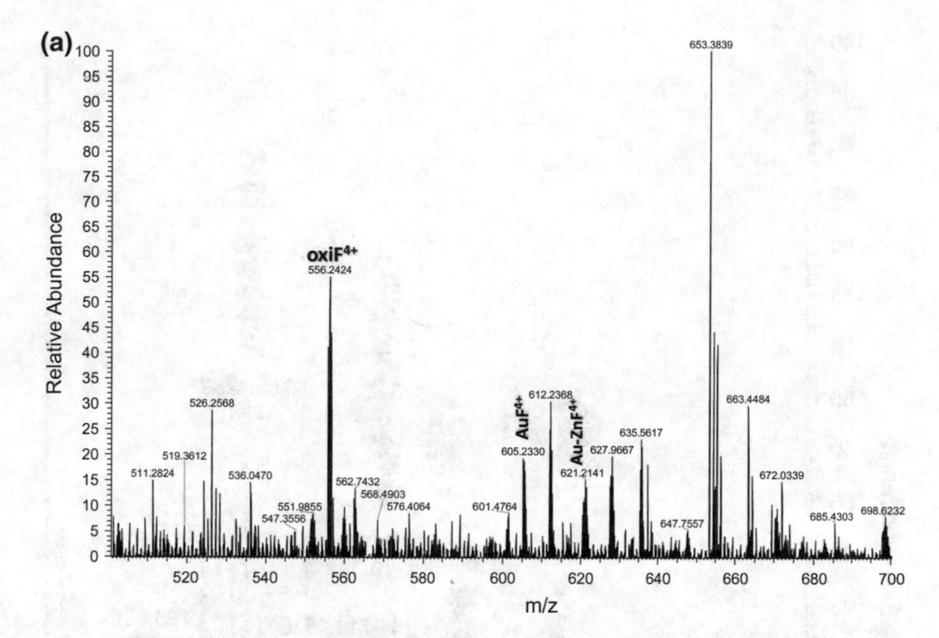

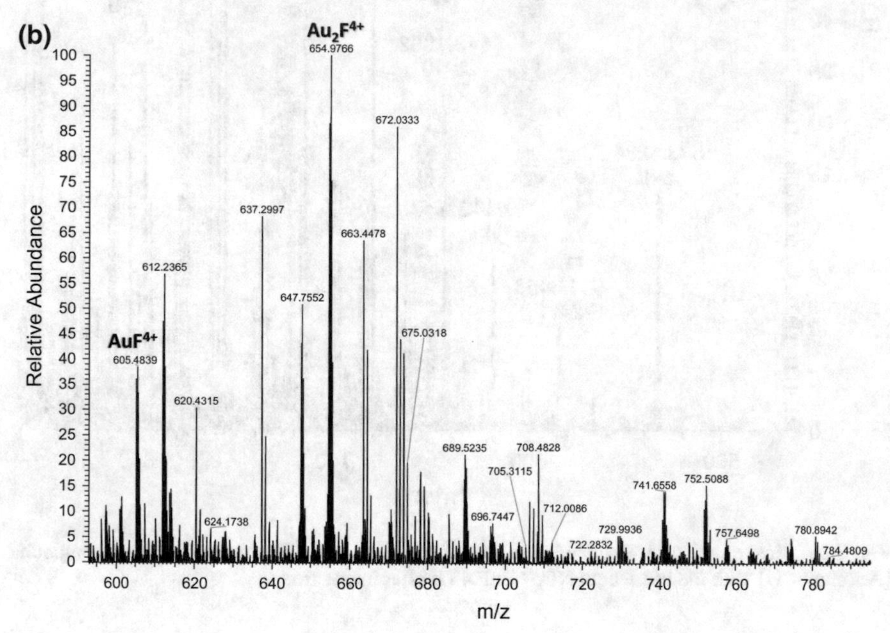

Fig. 5.27 Mass spectra obtained for the interaction of $[AuCl_2(phen)]^+$ with NCp7 (ZnF2) **a** immediately and **b** 48 h after mixing

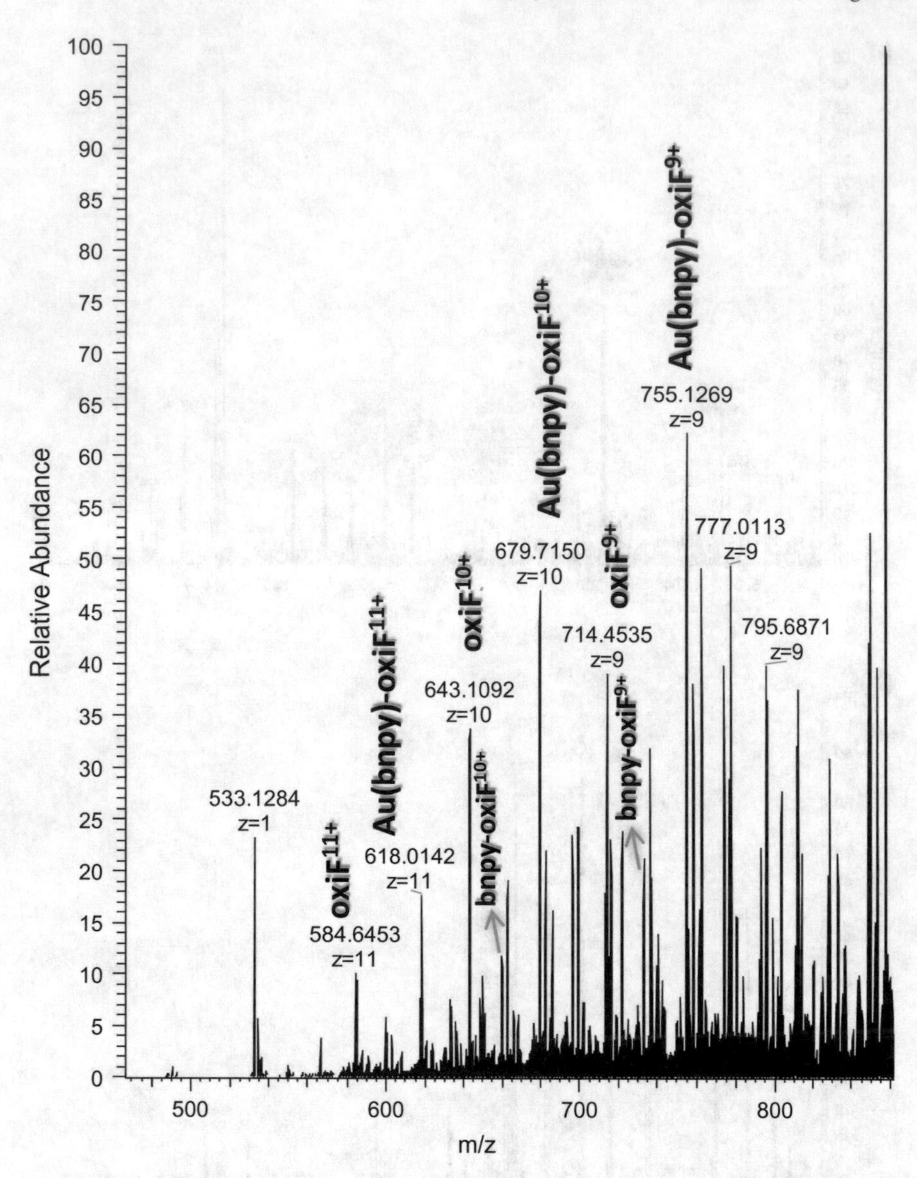

Fig. 5.28 Mass spectrum obtained for the interaction of the organometallic compound [Au(bnpy)Cl_2] with the full-length NCp7 ZnF 48 h after interaction

References

1. Rigobello, M.P., Scutari, G., Folda, A., Bindoli, A.: Mitochondrial thioredoxin reductase inhibition by gold(I) compounds and concurrent stimulation of permeability transition and release of cytochrome c. Biochem. Pharmacol. **67**(4), 689–696 (2004). https://doi.org/10.1016/j.bcp.2003.09.038
2. Henderson, W., Nicholson, B.K., Faville, S.J., Fan, D., Ranford, J.D.: Gold(III) thiosalicylate complexes containing cycloaurated 2-arylpyridine, 2-anilinopyridine and 2-benzylpyridine ligands. J. Organomet. Chem. **631**(1–2), 41–46 (2001). https://doi.org/10.1016/S0022-328X(01)00987-1
3. Janzen, D.E., Doherty, S.R., Vanderveer, D.G., Hinkle, L.M., Benefield, D.A.D.A., Vashi, H.M., Grant, G.J.: Cyclometallated gold(III) complexes with a trithiacrown ligand: solventless Au(III) cyclometallation, intramolecular gold? sulfur interactions, and fluxional behavior in 1,4,7-trithiacyclononane Au(III) complexes. J. Organomet. Chem. **755**, 47–57 (2014). https://doi.org/10.1016/j.jorganchem.2013.12.048
4. Pia Rigobello, M., Messori, L., Marcon, G., Agostina Cinellu, M., Bragadin, M., Folda, A., Scutari, G., Bindoli, A.: Gold complexes inhibit mitochondrial thioredoxin reductase: consequences on mitochondrial functions. J. Inorg. Biochem. **98**(10), 1634–1641 (2004). https://doi.org/10.1016/j.jinorgbio.2004.04.020
5. Parish, R.V., Howe, B.P., Wright, J.P., Mack, J., Pritchard, R.G., Buckley, R.G., Elsome, A.M., Fricker, S.P.: Chemical and biological studies of dichloro(2-((dimethylamino)methyl)phenyl)gold(III). Inorg. Chem. **35**(6), 1659–1666 (1996). https://doi.org/10.1021/IC950343B
6. Buckley, R.G., Elsome, A.M., Fricker, S.P., Henderson, G.R., Theobald, B.R.C., Parish, R.V., Howe, B.P., Kelland, L.R.: Antitumor properties of some 2-[(dimethylamino)methyl]phenylgold(III) complexes. J. Med. Chem. **39**(26), 5208–5214 (1996). https://doi.org/10.1021/jm9601563
7. Massai, L., Cirri, D., Michelucci, E., Bartoli, G., Guerri, A., Cinellu, M.A., Cocco, F., Gabbiani, C., Messori, L.: Organogold(III) compounds as experimental anticancer agents: chemical and biological profiles. Biometals **29**(5), 863–872 (2016). https://doi.org/10.1007/s10534-016-9957-x
8. Messori, L., Cinellu, M.A., Merlino, A.: Protein recognition of gold-based drugs: 3D structure of the complex formed when lysozyme reacts with Aubipy^{c}. ACS Med. Chem. Lett. **5**(10), 1110–1113 (2014). https://doi.org/10.1021/ml500231b
9. Spell, S.R., Mangrum, J.B., Peterson, E.J., Fabris, D., Ptak, R., Farrell, N.P.: Au(iii) compounds as HIV nucleocapsid protein (NCp7)–nucleic acid antagonists. Chem. Commun. **53**(1), 91–94 (2017). https://doi.org/10.1039/C6CC07970A
10. Jacques, A., Lebrun, C., Casini, A., Kieffer, I., Proux, O., Latour, J.-M., Sénèque, O.: Reactivity of Cys 4 zinc finger domains with gold(III) complexes: insights into the formation of "gold fingers". Inorg. Chem. **54**(8), 4104–4113 (2015). https://doi.org/10.1021/acs.inorgchem.5b00360
11. Cinellu, M.A., Zucca, A., Stoccoro, S., Minghetti, G., Manassero, M., Sansoni, M.: Synthesis and characterization of gold(III) adducts and cyclometallated derivatives with 2-substituted pyridines. Crystal structure of [Au{NC5H4(CMe2C6H4)-2}Cl2]. J. Chem. Soc. Dalt. Trans. (17), 2865–2872 (1995) https://doi.org/10.1039/dt9950002865
12. Casini, A., Diawara, M.C., Scopelliti, R., Zakeeruddin, S.M., Grätzel, M., Dyson, P.J., Abbott, B.J., Mayo, J.G., Shoemaker, R.H., Boyd, M.R.: Synthesis, characterisation and biological properties of gold(III) compounds with modified bipyridine and bipyridylamine ligands. Dalton Trans. **39**(9), 2239 (2010). https://doi.org/10.1039/b921019a
13. Sheldrick, G.M.: Crystal structure refinement with *SHELXL*. Acta Crystallogr. Sect. C Struct. Chem. **71**(1), 3–8 (2015). https://doi.org/10.1107/S2053229614024218
14. Dolomanov, O.V., Bourhis, L.J., Gildea, R.J., Howard, J.A.K., Puschmann, H.: OLEX2: a complete structure solution, refinement and analysis program. J. Appl. Crystallogr. **42**(2), 339–341 (2009). https://doi.org/10.1107/S0021889808042726

15. Bonamico, M., Dessy, G.: The crystal structure of anhydrous potassium tetrachloroaurate(III). Acta Crystallogr. Sect. B Struct. Crystallogr. Cryst. Chem. **29**(8), 1735–1736 (1973) https://doi.org/10.1107/s0567740873005406
16. Janiak, C.: A critical account on π–π stacking in metal complexes with aromatic nitrogen-containing ligands†. J. Chem. Soc. Dalt. Trans. (21), 3885–3896 (2000) https://doi.org/10.1039/b003010o
17. Chifotides, H.T., Dunbar, K.R.: Anion − π interactions in supramolecular architectures. Acc. Chem. Res. **46**(4), 894–906 (2013). https://doi.org/10.1021/ar300251k
18. Abbate, F., Orioli, P., Bruni, B., Marcon, G., Messori, L.: Crystal structure and solution chemistry of the cytotoxic complex 1,2-dichloro(o-phenanthroline)gold(III) chloride. Inorganica Chim. Acta **311**(1), 1–5 (2000). https://doi.org/10.1016/S0020-1693(00)00299-1
19. Sanna, G., Pilo, M.I., Spano, N., Minghetti, G., Cinellu, M.A., Zucca, A., Seeber, R.: Electrochemical behaviour of cyclometallated gold(III) complexes. Evidence of transcyclometallation in the fate of electroreduced species. J. Organomet. Chem. **622**(1), 47–53 (2001). https://doi.org/10.1016/S0022-328X(00)00822-6
20. Montero, D., Tachibana, C., Rahr Winther, J., Appenzeller-Herzog, C.: Intracellular glutathione pools are heterogeneously concentrated. Redox Biol. **1**(1), 508–513 (2013). https://doi.org/10.1016/j.redox.2013.10.005
21. Rychlewska, U., Warżajtis, B., Glišić, B.Đ., Živković, M.D., Rajković, S., Djuran, M.I.: Monocationic gold(III) Gly-l-His and l-Ala-l-His dipeptide complexes: crystal structures arising from solvent free and solvent-containing crystal formation and structural modifications tuned by counter-anions. Dalton Trans. **39**(38), 8906–8913 (2010). https://doi.org/10.1039/c0dt00163e

Chapter 6
"Dual-Probe" X-Ray Absorption Spectroscopy

6.1 Introduction

The success of antitumor Pt(II) compounds directs the scientific interest toward the isoelectronic Au(III) for the synthesis of therapeutic agents [1]. Although Au(III) is not very stable under biological conditions due to its proneness to reduction and fast ligand exchange rates, many recent Au(III)-containing compounds have been rationally designed by tailoring the ligands to stabilize Au(III), overcoming those limitations and still demonstrating relevant antitumor activities in vitro [2–4].

The Au(III) compounds evaluated so far as ZnF inhibitors often undergo reduction to Au(I) with loss of all ligands and concomitant electrophilic attack on the Zn-bound residues, resulting in zinc displacement [5, 6]. In the case of Au(III), it is commonly accepted that oxidation state of incorporated gold from a variety of complexes in {AuF} is +1, thus implicating concomitant oxidation of the peptide core. The use of N-donors, such as chelating ligands, nucleobases or simple planar amines in gold compounds have been suggested as potential stabilizers for the Au(III) oxidation state, even in the presence of peptides with high cysteine content such as ZnF. Farrell and co-workers developed a series of Au(III) compounds with tridentate diethylenetriamine, dien ($NH_2CH_2CH_2NHCH_2CH_2NH_2$, N^N^N coordination) where the fourth coordination site is occupied by a Cl^-, pyridine derivatives or nucleobases [7, 8]. Significative changes in the reactivity with ZnFs was observed for the $[Au(III)(dien)L]^{n+}$ motif depending on the nature of the ligand L.

However, the structural details of the products formed by the interaction of those Au(III) complexes with ZnFs after zinc displacement are currently lacking, hindering a mechanistic description of the Zn displacement process. This information is fun-

---

Parts of this chapter has been reproduced with permission from ACS. https://pubs.acs.org/doi/10.1021/acs.inorgchem.7b02406.

R. E. Ferraz de Paiva, *Gold(I,III) Complexes Designed for Selective Targeting and Inhibition of Zinc Finger Proteins*, Springer Theses,
https://doi.org/10.1007/978-3-030-00853-6_6

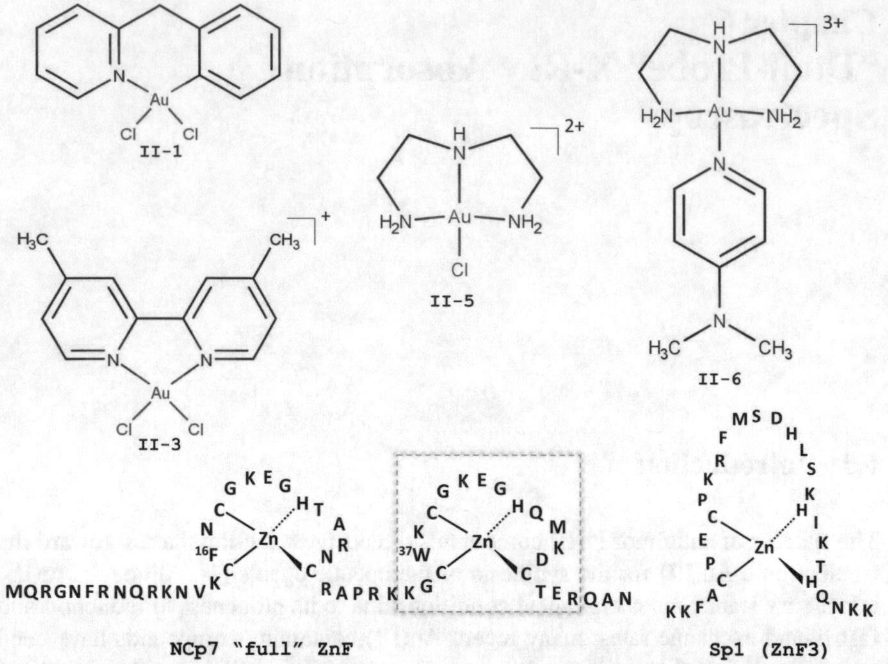

Fig. 6.1 Structural formulas of Au(III) compounds selected for this study. The zinc finger model peptides full-length NCp7 (F2 boxed) and Sp1 are also shown

damental for designing new compounds with better stability and specificity towards inhibition of the HIV-1 ZnF protein and also to obtain insights on the mechanism of inhibition. For that purpose, XAS represents a powerful technique for probing the oxidation state and coordination sphere of Au-containing species in solution. In addition, in the context of the interaction of metallodrugs with a metalloprotein, XAS can be used in a "dual-probe" approach, by monitoring both the absorption edge of the metal complex and also the edge of the metal present in the metalloprotein. As a proof-of-concept, we evaluated the interaction of Au(III) complexes with ZnFs by monitoring the Au L_3-edge and also the Zn K-edge. Furthermore, given the unique stability and reactivity of the Au(C^N) coordination motif discussed in Part II—Chap. 5, the interaction of the compound [Au(bnpy)Cl_2] with ZnFs was also studied by XAS. Figure 6.1 shows the Au(III) complexes evaluated here, along with the structures of the ZnF targets.

6.2 Experimental

6.2.1 *Synthesis and Zinc Finger Preparation*

H[$AuCl_4$] was acquired from Sigma-Aldrich and used without further purification. Complexes **II-1**, **II-3**, **II-5** and **II-6** were synthesized and purified according to published procedures [7–10]. Characterization of the synthesized compounds was performed by conventional spectroscopic techniques including ^1H, ^{13}C, ^{31}P NMR spectroscopy, mass spectrometry, elemental analysis, and infrared and UV spectroscopies, attesting the success of the synthetical procedures.

The zinc finger models used in this study were: HIV-1 nucleocapsid protein ZnF2 and "full" zinc finger (ZnF1 + ZnF2) and the human transcription factor Sp1 ZnF3. NCp7 ZnF2 and Sp1 ZnF3 were purchased from Aminotech Co. (São Paulo, Brazil), the full-length NCp7 ZnF was acquired from Invitrogen (USA). The apopeptides were checked by mass spectrometry and used as received. Both zinc fingers were prepared by dissolving sufficient mass of apopeptides in a 100 mmol L^{-1} solution of zinc acetate prepared in degassed water. The pH was adjusted to 7.2–7.4 using NH_4OH or HOAc if needed, leading to a solution with final concentration of 30 mmol L^{-1} of zinc finger. Sequences:

```
                        10         20         30         40         50
NCp7 ZnF2           KGCWKCGKEG HQMKNCTER
full-length NCp7    MQRGNFRNQR KNVKCFNCGK EGHTARNCRA PRKKGCWKCG KEGHQMKDCT ERQAN
Sp1 ZnF3            KKFACPECPK RFMSDHLSKH IKTHQNKK
```

6.2.2 *Sample Preparation*

Gold L_3-edge XAS measurements of the compound H[$AuCl_4$] were performed in solid state at XAFS1 beamline. The compounds [Au(bnpy)Cl_2] (**II-1**) and [$AuCl_2$(dmbipy)]$^+$ (**II-3**) were measured as solids at the XAFS 2 beamline. The solid samples used in the XAS measurements were finely grounded and diluted in boron nitride to a maximum X-ray absorbance of about 1. The homogeneous powder was then pressed into circular pellets of 13 mm diameter using a hydraulic press, placed in a plastic sample holder and covered with polyimide adhesive tape (Kapton) with about 40 μm thickness.

The complexes **II-5** and **II-6**, the isolated zinc fingers and the respective interaction products were measured in solution at the XAFS2 beamline [11]. For the interaction with ZnFs, stock solutions of **II-1**, **II-5** and **II-6** were prepared by dissolving the solid compounds in dimethylformamide (dmf) to a final concentration of 100 mmol L^{-1}. For evaluating the interaction of the model compounds with the zinc fingers, a total sample volume of 10 μL (30 mmol L^{-1}) was prepared by mixing sufficient volumes of the stock solutions of the Au(III) compounds **II-1, II-5** and **II-6** with stocks of the corresponding zinc finger in a molar ratio of 1:1. The interaction of compound **II-1** with N-acetyl-L-cysteine was also followed by XAS. For that purpose, a stock of **II-1** (83.3 mmol L^{-1} in dmso) was mixed with the proper amount of a stock of *N*-Ac-Cys (500 mmol L^{-1} in dmso) for a final sample concentration of 70 mmol L^{-1}. In each XAS measurement of the samples in solution, about 3 μL of the prepared solutions of the zinc fingers, Au(III) complexes in dmf or interaction products were placed in a plastic sample holder, covered with the same 40 μm thick Kapton adhesive and frozen in a closed-cycle liquid helium cryostat. The solution samples were kept below 50 K throughout the measurements.

Zinc K-edge XAS were also collected for the isolated zinc fingers and the interaction products of compound **II-5** and **II-6** with NCp7(ZnF2) and Sp1 (ZnF3). As comparison, the Zn K-edge XAS spectrum of the interaction of [AuCl(Et_3P)] with NCp7 (ZnF2) was also acquired. Zinc K-edge XAS was recorded in fluorescence mode at XAFS2 beamline using the same 15-element Ge solid-state detector and setting an integrating window of about 170 eV around the Zn $K\alpha_1$ and $K\alpha_2$ emission lines (8637.2 eV and 8614.1 eV, respectively). In the Zn K-edge XAS experiments the incoming energy was calibrated by setting the absorption edge of a Zn foil to 9659 eV.

Inspection of fast XANES scans revealed no signs of radiation damage for both Au L_3-edge and Zn K-edge measurements, and the first and last scans of each data set used in the averages were identical. Data averaging, background subtraction and normalization were done using standard procedures using the ATHENA package [12].

6.2.3 XAFS1 and XAFS2 Beamlines

Both beamlines are located at the Brazilian Synchrotron Light Laboratory (CNPEM/LNLS) [11, 13]. At the XAFS1 beamline the incident energy was selected by a channel-cut monochromator equipped with a Si(111) crystal, and at the XAFS2 beamline a double-crystal, fixed-exit monochromator was used. The beam size at the sample was approximately 2.5×0.5 mm^2 (horizontal × vertical) at XAFS1 and 0.4×0.4 mm^2 at XAFS2, with an estimated X-ray flux of 10^8 ph/s (XAFS1) and 10^9 ph/s (XAFS2). The incoming X-ray energy was calibrated by setting the maximum of the first derivative of L_3-edge of a gold metal foil to 11,919 eV. At XAFS1, XAS spectra were collected in conventional transmission mode using ion chambers filled with a mixture of He and N_2 while at XAFS2 fluorescence mode detection was used. The fluorescence signal was recorded using a 15-element Ge solid-state detector (model GL0055S—Canberra Inc.) by setting an integrating window of about 170 eV around the Au $L\alpha_1$ and $L\alpha_2$ emission lines (9713.3 eV and 9628.0 eV, respectively).

6.2.4 *TD-DFT*

TD-DFT calculations of the Au L_3-edges of compounds $[AuCl(dien)]^{2+}$ and $[Au(dien)(dmap)]^{3+}$ were used to make sure our protocol could reproduce the trends in energy shifts and intensities observed experimentally. TD-DFT-calculated L_3-edges have an overall good agreement with our experimental data, as shown in Fig. 6.7. All calculations were performed using the ORCA quantum chemistry code, version 3.0.3 [14]. All molecules were optimized at the PBE0/def2-TZVP level of theory using the def2-ECP effective core potential [15]. TD-DFT calculations (using the Tamm-Dancoff approximation) [16] as implemented in ORCA were performed with the PBE0 [17–19] functional using an all-electron scalar relativistic Douglas-Kroll-Hess Hamiltonian [20–22] with the DKH-def2-TZVP-SARC basis set [23]. The Au 2p to valence excitations were performed by only allowing excitations from the Au 2p donor orbitals to all possible virtual orbitals of the molecule (only limited by the number of calculated roots). Intensities include electric dipole, magnetic dipole and quadrupole contributions. The RIJCOSX approximation [24, 25] was used to speed up the Coulomb and Exchange integrals in both geometry optimizations and TD-DFT calculations.

6.2.5 *Mass Spectrometry*

A stock solution of the $[Au(dien)(dmap)]^{3+}$ in acetonitrile was mixed with a stock solution of the full-length NCp7 ZnF in water (1:1 mol/mol of Au complex per ZnF core). The pH final solution was adjusted to 7.2 with NH_4OH. leading to a solution containing the interaction product in a 1 mmol L^{-1} concentration, which was incubated for up to 24 h at room temperature. MS experiments were carried out on an Orbitrap Velos from Thermo Electron Corporation operated in positive mode. Samples (25 μL) were diluted with methanol (225 μL) and directly infused at a flow rate of 0.7 μL/min using a source voltage of 2.30 kV. The source temperature was maintained at 230 °C throughout the experiment.

6.3 Results and Discussion

The Au(III) compounds $[AuCl_4]^-$, $[Au(bnpy)Cl_2]$ (**II-1**), $[AuCl_2(dmbipy)]^+$ (**II-3**), $[AuCl(dien)]^{2+}$ (**II-5**) and $[Au(dien)(dmap)]^{3+}$ (**II-6**) were selected as experimental models. XAS was used to evaluate the interaction of **II-1**, **II-5** and **II-6** with NCp7 (ZnF2). Compounds **II-5** and **II-6** were evaluated when interacting with Sp1 (ZnF3).

A detailed analysis of the XAS spectra of the compounds studies here is given in Appendix of this chapter, including a summary of the spectroscopic features observed for each compound (Fig. 6.6 and Table 6.1). Figure 6.7 shows the TD-DFT spectra

Table 6.1 Edge position, oxidation state and approximated site symmetry from studied compounds and gold(0) reference

Compound		Edge position (eV)	Oxid. St.	Site symmetry (around Au atom)
#	Name			
II-1	$[Au(bnpy)Cl_2]$	11,918.9	3+	C_1
II-3	$[AuCl_2(dmbipy)]^{1+}$	11918.5	3+	C_{2v}
	$[AuCl_4]^-$	11917.6	3+	D_{4h}
II-5	$[AuCl(dien)]^{2+}$	11918.1	3+	C_{2v}
II-6	$[Au(dien)(dmap)]^{3+}$	11918.5	3+	C_{2v}
Reaction	Name	Edge position (eV)	Oxid. St.	Coordination sphere
II-1+NCp7	$[Au(bnpy)Cl_2]$+NCp7 ZnF2	11919.7	1+	L-Au-L
II-1+*N*-Ac-Cys	$[Au(bnpy)Cl_2]$+ *N*-Ac-Cys	11919.9	3+	(bnpy-κ^2C,N)-Au-S
II-5+NCp7	$[AuCl(dien)]^{2+}$ +NCp7 ZnF2	11919.7	1+	(Cys)S-Au-N(His)
II-6+NCp7	$[Au(dien)(dmap)]^{3+}$ + NCp7 ZnF2	11918.9	3+	(dien-κ^3N,N',N'')-Au-N
II-5+Sp1	$[AuCl(dien)]^{2+}$ +Sp1 ZnF3	11919.2	1+	(His)N-Au-N(His)
II-6+Sp1	$[Au(dien)(dmap)]^{3+}$ + Sp1 ZnF3	11919.8	1+	(His)N-Au-N(His)

obtained for compounds **II-5** and **II-6**. It is important to note that the reaction products discussed here were obtained and immediately frozen for the spectroscopic measurements (unless otherwise stated). Therefore the spectra obtained and discussed in this chapter represent a snapshot of the species that appear at the very beginning of the reactions evaluated.

6.3.1 Au L_3-Edge: The Au(C^N) Versus the Au(N^N) Motif

XAS has the major advantage of being sensitive to both the vicinity (nature of the ligands) and the oxidation state of the metal center. The interaction of the organometallic compound $[Au(bnpy)Cl_2]$ with NCp7 ZnF2 was studied in comparison to $[AuCl_2(dmbipy)]^+$ by XAS. When comparing the spectra obtained for the two model compounds, it becomes clear that the N^N bidentate ligand leads to a Au L_3-edge spectrum that better resembles that of the spectra discussed in the previous sub-session for $[Au(dien)L]^n$ complexes, with a strong white line peak

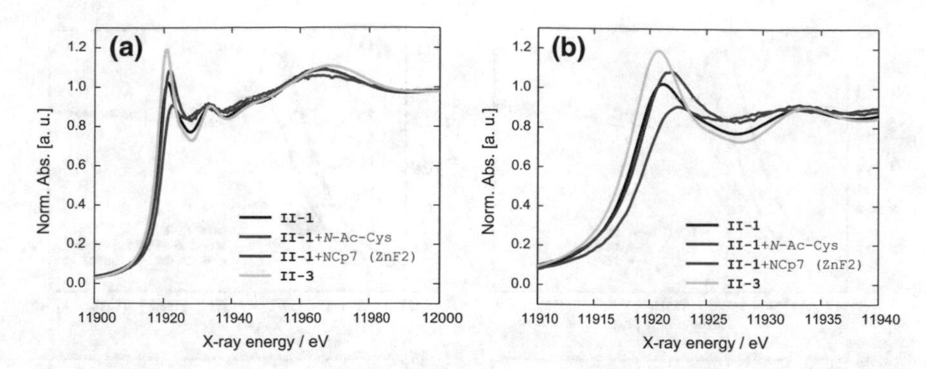

Fig. 6.2 **a** XANES spectrum of compound **II-1** and interactions with NCp7 and model S-donor *N*-Ac-Cys. A selected typical Au(N^N) compound (**II-3**) is also shown for comparison. **b** expansion around the white line

(normalized intensity of ~1.2 a.u.). On the other hand, the organometallic compound [Au(bnpy)Cl_2] had a much weaker white line signal, indicating fewer *d* holes as consequence of deprotonated bnpy acting as a better electron donor than dmbipy.

The interactions of compound [Au(bnpy)Cl_2] with the model S-donor *N*-Ac-Cys and with NCp7 (ZnF2) were also studied by XAS. Interestingly, two different behaviors were observed.

An increase in intensity of the white line is observed for the interaction [Au(bnpy)Cl_2] + *N*-Ac-Cys. That observation alone is a strong evidence that a ligand replacement reaction took place, with one of the chlorides found in the organometallic compound being replaced by *N*-Ac-Cys. The thiolate has stronger donating properties when compared to chloride, which explains the increase in intensity observed here.

On the other hand, a prominent decrease of the white line intensity was observed for the interaction [Au(bnpy)Cl_2] + NCp7 (ZnF2). The L_3-edge spectrum obtained here resembles, but is not identical to, the theoretical model Cys-Au(I)-His (see Part I—Chap. 3, Fig. 3.3). The similarity between the Au-L_3 XANES spectrum obtained for **II-1**+NCp7 and the TD-DFT calculated spectrum of theoretical model **T-4** support the hypothesis that, after the reductive elimination step that leads to the C-S coupling (see Scheme 5.1, Part II—Chap. 5), Au(I) stays coordinated to the pyridine ring found in the bnpy moiety. Despite the difference in the ring size, a pyridine (from bnpy) or a His residue coordinated to Au(I) are expected to have similar XANES spectra (Fig. 6.2).

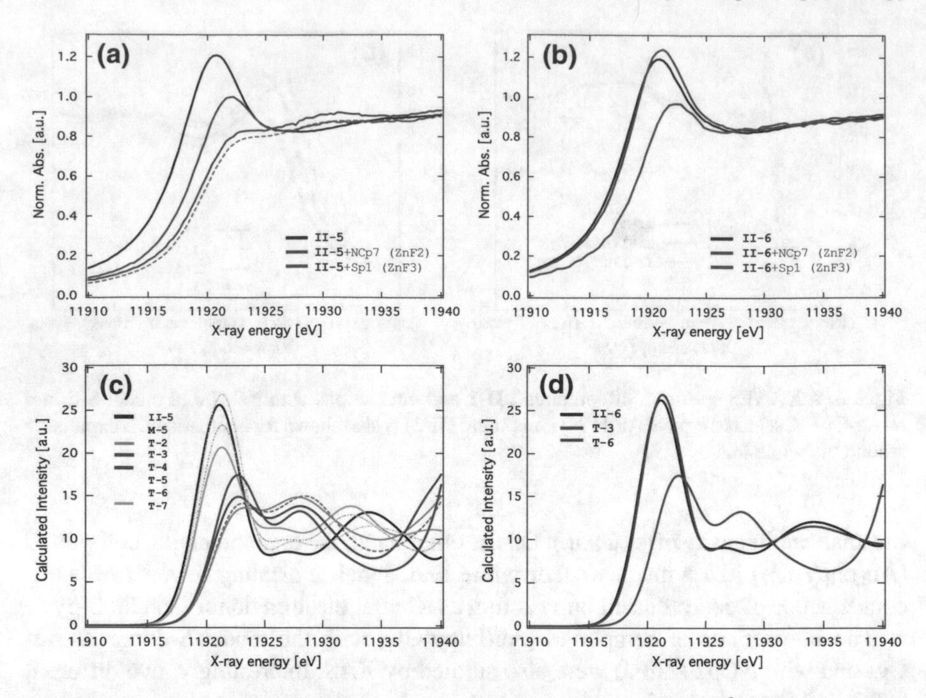

Fig. 6.3 Gold L_3-edge XANES spectra of **a** compound **II-5** ($[AuCl(dien)^{2+}]$) and its respective reaction products with NCp7 and Sp1 (red and blue, respectively), **b** compound **II-6** ($[Au(dien)(dmap)]^{3+}$) and its respective reaction products with NCp7 and Sp1 (red and blue, respectively). The spectrum of model compound [Au(*N*-Ac-Cys)] (**M-5**) is also shown (gray dashed line) for comparison. **c** and **d** TD-DFT-calculated XANES spectra of the Au(III) compounds **II-5** and **II-6** together with the theoretical models for the coordination sphere of the products obtained upon interaction of Au(III) complexes with NCp7 (theoretical models Cys-Au(I)-Cys (**T-2**) and His-Au(I)-Cys (**T-4**)) and Sp1 (theoretical models His-Au(I)-His (**T-3**), dien-Au(III)-Cys (**T-5**), dien-Au(III)-His (**T-6**) and Au(III)-Cys_2-His_2 (**T-7**)). The optimized structures of the Au(III) theoretical models are shown in Fig. 6.4. The energy shift of 465 eV was applied to the calculated spectra

6.3.2 *Au L_3-Edge: Reactions of Au(N^N^N) Complexes with ZnF Proteins*

The XANES spectra of the Au(III) model compounds **II-5** and **II-6** are presented in Fig. 6.3, together with the reaction products formed after their interaction with NCp7 and Sp1 zinc fingers. As expected for Au(III) compounds, these spectra contain a strong absorption in the white line region as a consequence of empty Au d orbitals promoting the allowed $2p_{3/2} \rightarrow 5d$ transitions in the L_3-edge XAS. Compounds **II-2** and **II-3** have a white line peak located at about 11,921 eV with a normalized peak intensity of about 1.2 units.

NCp7. The interaction product **II-5**+NCp7 presents a white line of low intensity, similar to that expected for the previously examined Au(I) compounds or an even more reduced gold species, thus suggesting that a reduction of Au(III) took place upon the interaction of **II-5** with NCp7. Structurally, **II-5**+NCp7 differs from the pure compound $[AuCl(dien)]^{2+}$, as evidenced by both the XANES (Fig. 6.3a) and EXAFS (Fig. 6.8).The XANES of product **II-5**+NCp7 closely resembles that of model compound **M-5** from Part I—Chap. 3 (coordination sphere model shown in Fig. 6.9), presenting a rather weak white line peak, a pronounced post-edge feature at around 11,930 eV and another broad one at 11,940–11,985 eV (Fig. 6.3a). Moreover, the k^2-weighted EXAFS spectra of **II-5**+NCp7 and that of model compound **M-5** are virtually superimposable up to $k = 10$ Å^{-1}, indicating that both contain a similar environment around the gold atom, i.e., linear S-Au-S coordination with a similar Au-S distance of about 2.30 Å (Fig. 6.8). This corroborates the hypothesis that upon reaction with NCp7, $[AuCl(dien)]^{2+}$ loses all its ligands and undergoes Au(III) $\rightarrow$ Au(I) reduction, leading to the formation of gold finger in a Cys-Au(I)-Cys linear geometry, the same coordination sphere observed for other Au-protein adducts reported elsewhere [26, 27]. Additionally, it also agrees with previous mass spectrometry studies showing the formation of only gold finger (without ligands) when $[AuCl(dien)]^{2+}$ interacts with NCp7 [6, 8].

In contrast, the spectra of reaction product **II-6**+NCp7 (Fig. 6.3b) resembles that of its precursor, $[Au(dien)(dmap)]^{3+}$, with only a slight increase in the white line intensity. Additionally, despite the relatively low quality, the EXAFS data overlaps up to $k = 8$ Å^{-1}, which is sufficient to conclude that they have identical first coordination shell around the gold (Fig. 6.10). These results indicate that the Au(III) in $[Au(dien)(dmap)]^{3+}$ remains mostly unreacted upon interaction with NCp7, on contrary to what is observed in the case of **II-5**. This is in agreement with MS data reported previously, which shows zinc displacement (apopeptide signal) and signals of $[Au(dien)dmap]^{3+}$. Thus, to some extent, Au(III) redox stability was achieved, even in the presence of the rich Cys content of NCp7, by modulating the first coordination sphere of Au(III)-dien complex. However, the observed changes in the white line intensity indicate different degrees of reduction after interaction of **II-5** and **II-6** with NCp7 (Fig. 6.11). In the case of **II-5**+NCp7, the pronounced diminishing of the white line points to a more reduced gold metal center when compared to **II-6**+NCp7.

Sp1. **II-5**+Sp1 and **II-6**+Sp1 still present a well-defined white line peak and the obtained XANES spectra are identical, with the white line peak shifting slightly to higher energies (about 1.4 eV) and reducing its intensity to about 1.0 normalized units, suggesting a change in oxidation state.

TD-DFT. TD-DFT-calculated L-edge spectra of different models were again fundamental in revealing the nature of the interaction products. Figure 6.3c, d shows the TD-DFT spectra of compounds [Au(*N*-Ac-Cys)] (**M-5** from Part I—Chap. 3), **II-5**, **II-6** and theoretical models for **II-5**+NCp7, **II-6**+NCp7, **II-5**+Sp1 and **II-6**+Sp1. The DFT-optimized structures of the theoretical Au(III) models, supplementing the already considered Au(I) models (Part I—Chap. 3, Figure 3.3) are presented in Fig. 6.4 (**T-5**: (dien)Au(III)-Cys, **T-6**: (dien)Au(III)-His and **T-7**: Cys_2-Au(III)-His_2).

The products **II-5**+Sp1 and **II-6**+Sp1 have almost identical XAS spectra, pointing to the formation of identical species (Figs. 6.3b and 6.11). Figure 6.6 shows that these species are, however, different from the purified gold finger, **II-5**+Sp1(AuF).

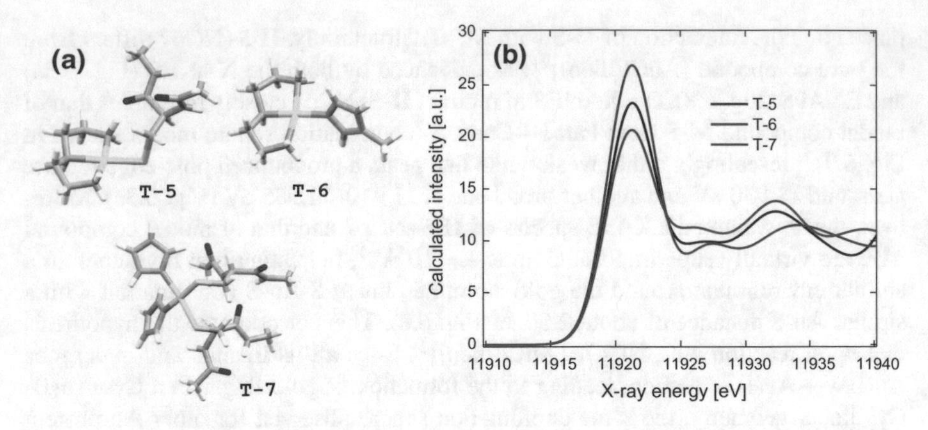

Fig. 6.4 **a** DFT optimized structures of the theoretical Au(III) models proposed for the reaction products of **II-5** and **II-6** with the ZnF proteins. Cys was modeled as *N*-Ac-L-Cys and His residues were modeled as 5-methylimidazole or imidazole. **b** TD-DFT-calculated spectra of the theoretical Au(III) models

The EXAFS spectra of these two reaction products with Sp1 presented rather poor quality, which prevented a complete structural analysis. This was due to the limited achievable concentration and sample quantity, which was more critical for **II-6**+Sp1. Reaction products **II-5**+Sp1 and **II-6**+Sp1 show experimentally a reduction in white line peak intensity and a slight energy shift and appear to result in identical products. For comparison with the experimental spectra obtained for **II-5** + Sp1 and **II-6** + Sp1, the theoretical models **T-3** (from Part I—Chap. 3) and **T-5** to **T-7** were considered. The Au(III) theoretical models do not reproduce the experimental trends observed. For (dien)Au(III)-His (**T-6**), a white line with higher intensity than observed for compounds **II-5** and **II-6** was obtained. The calculated spectrum of model (dien)Au(III)-Cys (**T-5**) shows a rather small reduction in the peak intensity, inconsistent with the *ca.* 20% reduction observed experimentally. The spectrum of the theoretical model Cys_2-Au(III)-His_2 (**T-7**) reproduces the observed reduction in intensity. However it does not account for the energy shift observed experimentally, which is likely a result of Au(III) $\rightarrow$ Au(I) reduction. The His-Au(I)-His model assumes that the two Cys residues in the Sp1 zinc finger are oxidized to form a disulfide bond and this electronic transfer reduces Au(III) to Au(I), resulting in ligand loss and binding to the two remaining His residues to Au(I). The calculated XANES spectrum for theoretical model **T-3** accounts for both intensity reduction with respect to the original compounds **II-5** and **II-6**, and the correct energy shift in the white line peak. Moreover, the second feature at about 11,928 eV is also present in the experimental XAS data of **II-5**+Sp1 and **II-6**+Sp1, being slightly broader.

For **II-5**+NCp7, the TD-DFT calculations show that both theoretical models Cys-Au(I)-Cys (theoretical model **T-2** from Part I—Chap. 3) and His-Au(I)-Cys (**T-4**, also from Part I—Chap. 3) reproduce the general behavior observed experimentally, i.e., a diminishing of the white line intensity and an energy shift of about 1.7 eV in the peak position when compared to the spectrum of intact $[AuCl(dien)]^{2+}$. Both theoretical models indicate that loss of all ligands, Au(III) → Au(I) reduction and coordination to either two cysteine residues or one cysteine and one histidine have occurred upon interaction of $[AuCl(dien)]^{2+}$ with NCp7. As discussed for **II-5**+Sp1 and **II-6**+Sp1, if the Au(III) center is reduced, it is likely that two Cys residues will form a disulfide bond (which gives off 2 electrons, reducing Au(III) to Au(I)). In the NCp7 case, the two remaining residues are one Cys and one His, suggesting that His-Au(I)-Cys is the most likely model. In fact, the calculated spectrum of the His-Au(I)-Cys model gives a slightly higher white line peak intensity than compound [Au(*N*-Ac-Cys)] (**M-5**), which is consistent with the experimental observation of **II-5**+NCp7 having slightly higher intensity than compound [Au(*N*-Ac-Cys)].

Compound **II-6** differs from the other Au(III) model compounds in the sense that it retains its oxidation state and the square-planar geometry upon interaction with NCp7. Moreover, MS studies show that the Zn(II) ion is indeed ejected in the interaction product **II-6**+NCp7 [8]. This result suggests two possibilities: one is the non-covalent interaction between $[Au(dien)(dmap)]^{3+}$ and NCp7, with **II-6** behaving as the class of ZnF inhibitors proposed by Garg and Torbett [28], with a mechanism of zinc displacement other than the electrophilic attack observed for most Au(I,III) compounds. Since $[Au(dien)(dmap)]^{3+}$ is a potent π-stacker (as indicated by tryptophan fluorescence quenching), we can hypothesize that compound **II-6** may disrupt the structure of the target ZnF, causing zinc displacement by π-stacking with neighbor aromatic residues (Phe16 and Try37). We were also able to identify the non-covalent adduct between compound **II-6** with the dinuclear NCp7 complete ZnF by MS (Fig. 6.12), further supporting the importance of an initially π-stacked species in the recognition of $[Au(dien)(dmap)]^{3+}$ by NCp7. The second hypothesis is the selective electrophilic attack on the histidine residue followed by replacement of the dmap ligand and maintenance of the (dien)Au-N coordination sphere. The possible mechanisms of zinc displacement caused by compound **II-6** are shown in Scheme 6.1.

We recently demonstrated that $[Au(dien)(9\text{-ethylguanine})]^{3+}$, an analog of compound **II-6** where dmap is replaced by the nucleobase 9-ethylguanine, is capable of binding to NCp7 dinuclear “full” zinc finger as demonstrated by MS and circular dichroism, with zinc displacement and gold incorporation into the protein structure with loss of all ligands [29]. $[Au(dien)(9\text{-ethylguanine})]^{3+}$ is also capable of inhibiting the interaction of the full-length NCp7 zinc finger with a small model for its natural substrate, SL2 RNA. The inhibition was demonstrated by MS, gel shift assay and fluorescence polarization, all of which show the release of SL2 DNA from the full-length NCp7 zinc finger once the system is treated with the Au(III) compound. Here we demonstrated that compound **II-6** was even more redox stable than $[Au(dien)(9\text{-ethylguanine})]^{3+}$, and our previous MS works [8] also indicate that **II-6** remains stable after interaction with NCp7 (ZnF2), with no signs of AuF being formed. Furthermore, compound **II-6** was also assayed as an inhibitor of the full-length NCp7 ZnF—SL2 RNA interaction by fluorescence polarization and was shown to be a potent inhibitor ($IC_{50} = 29\ \mu mol\ L^{-1}$). Combined with the XAS data discussed in this Chapter, the unique behavior of **II-6** points towards Zn displace-

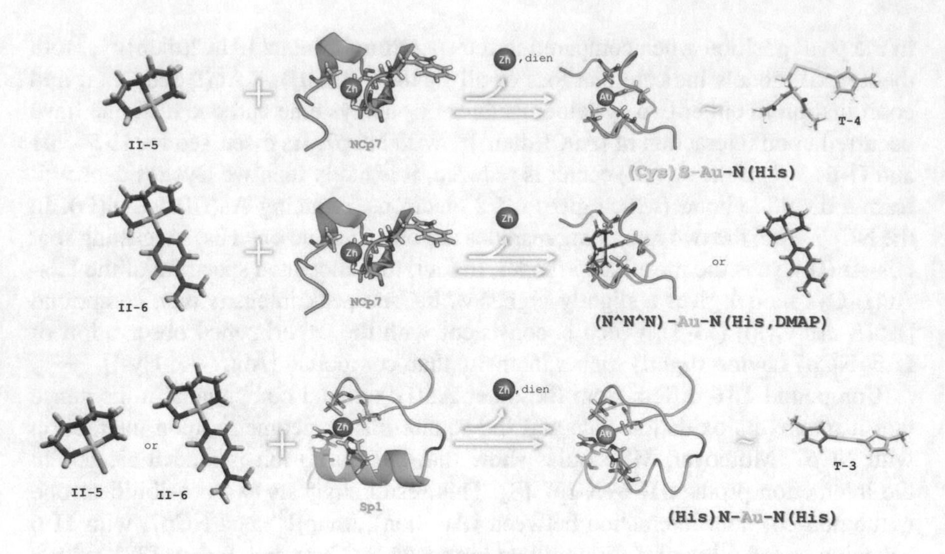

Scheme 6.1 Proposed mechanisms of Zn(II) displacement caused by $[Au(dien)Cl]^{2+}$ (**II-5**) and $[Au(dien)(dmap)]^{3+}$ (**II-6**) upon interaction with NCp7 and Sp1. The final coordination sphere of Au is highlighted in each case

ment and inhibition of full-length NCp7 ZnF—SL2 RNA interaction based on a non-covalent mechanism.

Comparing the final species formed for interaction of Au(III) compound (Scheme 6.1) with those formed by reacting Au(I) compounds (Part I—Chap. 3, Scheme 3.1) with the selected ZnF protein, it is possible to observe that they are not the same. Reduction versus oxidation is a hot topic in Au chemistry and it has been studied extensively by van Eldilk [30, 31]. In general, for Au(III) complexes reduction is really fast, to the point that ligand replacement cannot be observed experimentally. On the other hand, with the proper ligands (chelators, strong σ-donors), the redox process can be slowed down and ligand replacement will start to be relevant. For the Au(III) compounds studied here (**II-5** and **II-6**), the fact that the final interaction products with NCp7 and Sp1 are not the same observed for the Au(I) compounds is indicative of an important ligand-replacement step. Furthermore, the harder Lewis acid character of Au(III) versus Au(I) can play an important role in the His selectivity observed for Au(III) compounds.

6.3.3 Zinc K-Edge

The zinc K-edge XAS can be a powerful tool to get insights about the mechanism of zinc displacement from ZnFs, as differences in geometry and coordination sphere arise as consequence of the interaction with Zn ejectors, including the gold complexes

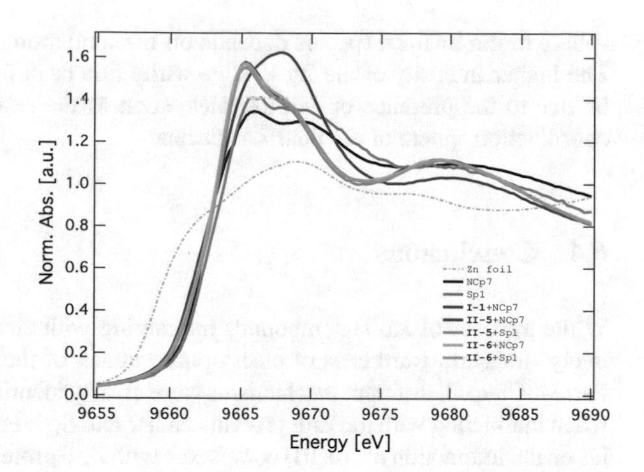

Fig. 6.5 Zinc K-edge XANES spectra of the pure NCp7 (ZnF2) and Sp1 (ZnF3) proteins, and the interaction products with Au(I) compound [AuCl(Et_3P)] and Au(III) compounds **2** and **3**

evaluated here. K-edge EXAFS has been previously used to structurally characterize zinc finger binding sites in NCp7 [32]. The Zn K-edge XANES of the free zinc finger proteins NCp7 and Sp1 are shown in Fig. 6.5, together with those from the reaction products obtained from the interaction with the Au(I) and Au(III) model compounds [AuCl(Et_3P)] (**I-1**, from Part I—Chap. 3), **II-5** and **II-6**, respectively. In the case of Au(I), only data of the reaction with [AuCl(Et_3P)] + NCp7 was acquired. In all cases, the EXAFS data quality does not allow a complete structural determination of the ejected zinc species.

A correlation between white line intensity and the coordination number of zinc in a metalloprotein is well documented: the white line intensity increases with the coordination number [33, 34]. Furthermore, it has been noted that the presence of Cys residues in the first coordination sphere of Zn diminishes the intensity of the white line, while the presence of carboxylate ligands such as Glu and Asp lead to an increase in intensity [34]. A theoretical and experimental trend was established by Ginchini et al. [33], which says that white line intensities <1.5 correspond to coordination numbers of 3 and 4, while a white line intensity of 1.6 and higher correspond to coordination numbers of 5 and 6.

Comparing our data to coordination number and donor atoms trends described above, we can see that the white line intensity of Sp1 (ZnF3) is indeed slightly higher than that of NCp7 (ZnF2), as consequence of the lower Cys count present in the coordination sphere of Zn(II) in Sp1. As for the interaction products, we can see that all Au + ZnF products led to a Zn K-edge white line intensity of 1.6, while the free ZnFs have white line intensities of 1.4 or lower. This increase suggests an expansion of the coordination number of zinc upon interaction with the target protein. Regarding the first coordination sphere of Zn after interaction with the ZnFs, we can see two different behaviors depending of the oxidation state of the Au complexes tested. Au(III) complexes led to spectra with two features in the XANES region, while the Au(I) complex [AuCl(Et_3P)] led to a spectrum with a broad and slightly more intense signal in the near-edge region. That is a clear evidence that the coordination sphere in the final Zn species depends on the oxidation state of the Au compound. The higher intensity of the Zn K-edge white line peak found in the Au(I) case can be due to the presence of carboxylates, such as the residues Glu and Asp, in the coordination sphere of the final Zn species.

6.4 Conclusions

While the case of Au(I) compounds interacting with zinc fingers represents a relatively straightforward case of electrophilic attack of the zinc core, as discussed in Part I—Chap. 3, different mechanisms were clearly identified for Au(III) compounds when interacting with the ZnF proteins NCP7 and Sp1. Previous spectroscopic studies on the interaction of Au(III) complexes with ZnF proteins indicated final products being pure Au(I) species with loss of all ancillary ligands (dien, bipyridine and others). However, we have already observed different degrees of reduction of gold and maintenance of (most of) the coordination environment in some systems previously investigated [5, 8, 35]. The XAS experiments, combined with TD-DFT calculations, confirm the ability to modulate the Au(III)-ZnF reaction by a suitable choice of ligands. Moreover, it is presented for the first time, structural-sensitive information on these reaction products.

Specifically, we showed that compound **II-1** suffers different degrees of reduction when interacting with the model S-donor *N*-Ac-Cys or with NCp7. With *N*-Ac-Cys, the interaction product has Au(III) in a coordination sphere consistent with replacement of the chlorides by a better donor, thiolate. On the other hand, when interacting with NCp7 a mixed population of Au(I,III) was observed, possibly as consequence of reductive elimination step that leads to a C-S coupling between the ligand bnpy and a Cys residue from the protein.

Regarding the Au(dien) complexes, **II-5** ($AuClN_3$ coordination sphere) reacts promptly with NCp7 (Cys_3His) and Sp1(Cys_2His_2) producing interaction products with different degrees of reduction in the gold site and different final geometries. The reduction in the white line intensity observed from the pure compound **II-5**, to the interaction **II-5**+Sp1 and finally **II-5**+NCp7, reflects the respective decrease in the number of 5d empty orbitals in this series and thus an increased reduction in the gold center. The final compounds were found to change from square-planar to linear geometry, with Cys-Au-Cys and His-Au-His coordination in the case of interaction of **II-5** with NCp7 and Sp1, respectively. In contrast, compound **II-6** (AuN_4 coordination sphere) is found to initially maintain the oxidation state and coordination geometry after zinc displacement in NCp7, with a final (dien)Au-N, coordination sphere where the N-donor is dmap or a histidine residue from the protein. To our knowledge, this is the first example of stabilization of a square-planar Au(III)-ligand moiety on a ZnF. It suggests an interaction relying on a mechanism other than a direct electrophilic attack as commonly observed in Au(I) compounds, with the initial gold species possibly being stacked with the tryptophan residue in NCp7. Furthermore, upon reaction with Sp1 compound **II-6** undergoes a marked

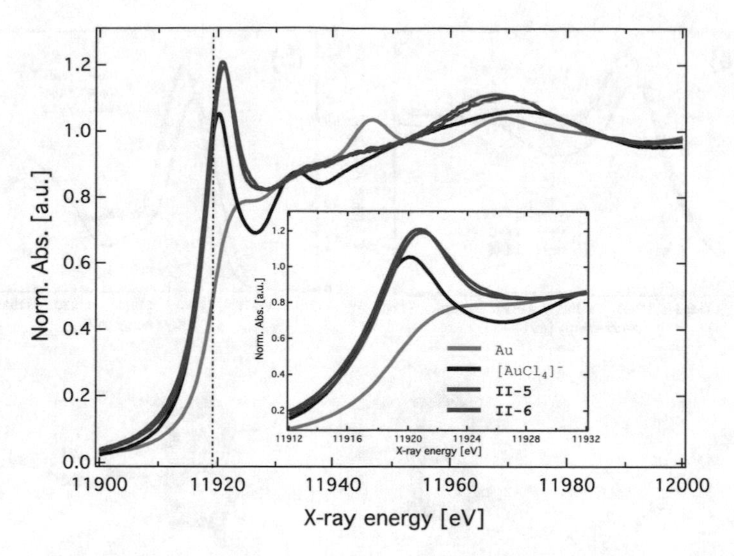

Fig. 6.6 Gold L_3 XANES spectra of the Au(III) complexes evaluated here. The spectrum obtained for a Au(0) foil is also shown and the vertical line indicates its edge position (19,919 eV)

reduction of the Au center, resulting in a final product identical to **II-5**+Sp1, i.e., His-Au-His in linear geometry. This suggests that the probable mechanism for the integration of Sp1 with gold(III) compounds is the usual electrophilic attack to the cysteines, causing zinc displacement and followed by reduction of gold (possibly via oxidative disulfide bond formation). These differences allow specific ZnF targeting considering the ZnF core intrinsic reactivity and also the gold coordination sphere. The Zn K-edge XAS spectra of the free ZnFs confirmed the trend between higher white line intensities and lower Cys count in the coordination sphere of Zn. Regarding the interaction products, zinc displacement is confirmed for both Au(I) and Au(III) cases, with expansion of the coordination number from 4 in the free ZnFs to 5 or 6 upon interaction with the Au complexes. The coordination sphere of the displaced Zn species depends on the oxidation state of the Au. Further studies are still needed to provide a detailed picture of the structure of the final Zn(II) species.

Appendix

XAS of Au(III) Model Compounds and TD-DFT Calculations

The spectra in Fig. 6.1 show distinguishing features in the white line region, which originate from dipole-allowed $2p_{3/2} \rightarrow 5d$ transitions, containing both metal-centered $5d_{3/2}$ and $5d_{5/2}$ final states (Fig. 6.6).

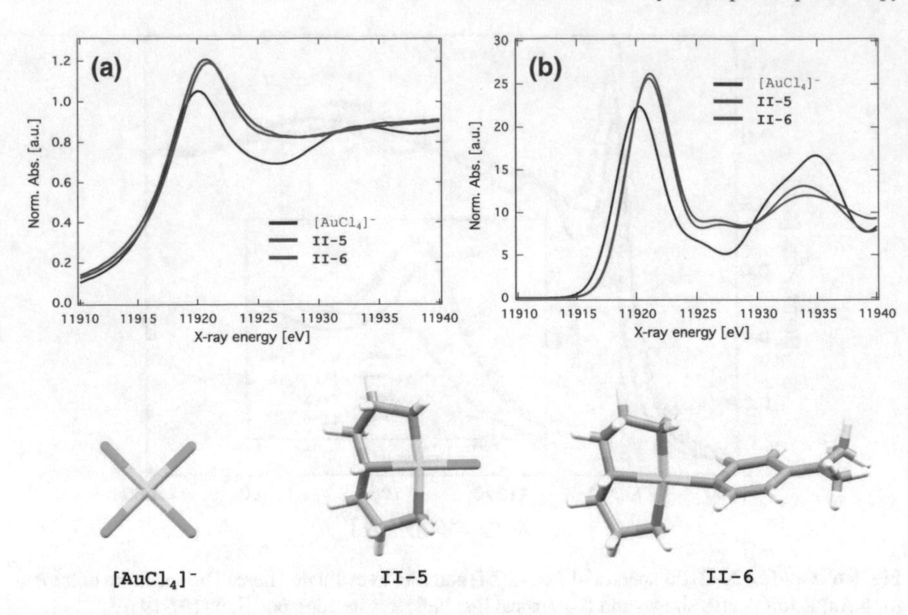

Fig. 6.7 **a** Experimental Au L_3-edge spectra Au(III) compounds **II-5** and **II-6** in comparison to $[AuCl_4]^-$. **b** TD-DFT-calculated Au L_3-edge spectra, shifted by 465 eV to lower energies. The DFT-optimized structures of teh experimental model compounds are also shown (bottom)

The Au(III) compounds ($[5d^8 6 s^0]$ electronic configuration) have white line peaks with high intensity (Fig. 6.6) in comparison to the Au(I) model compounds discussed earlier (Part I—Chap. 3, Fig. 3.5), as consequence of empty d orbitals promoting the allowed $2p_{3/2} \rightarrow 5d$ transitions in the L_3-edge XAS. In principle, the intensity of the white line can be used as a fingerprint of the d-electron count in these cases; however, in some cases the standard relationships between experimental Au L_3-edge white line intensities and oxidation state does not hold [36]. The XANES spectrum of compound $[AuCl_4]^-$ contains a sharp peak in the white line region and presents a distinct XANES spectrum when compared to compounds **II-5** and **II-6**, $[AuCl(dien)]^{2+}$ and $[Au(dien)(dmap)]^{3+}$, respectively. Compounds **II-5** and **II-6** have their white line maxima shifted by 0.8 and 1.0 eV respectively, in comparison to compound $[AuCl_4]^-$, indicating that the *dien* ligand has an oxidizing effect on the Au center (higher count of d holes on Au). That suggests a higher stability of the Au(III) species bound to chelating N- donor ligands, which directly translates into higher stability under biological reducing media. Point symmetry also contributes to the intensity of the white line in the gold L_3 XAS. When coordinated, *dien* leads to non-centrosymmetric groups. The p-d hybridization is enhanced reducing the effective d electron count in gold, thus increasing the intensity of the white line (Fig. 6.7 and Table 6.1).

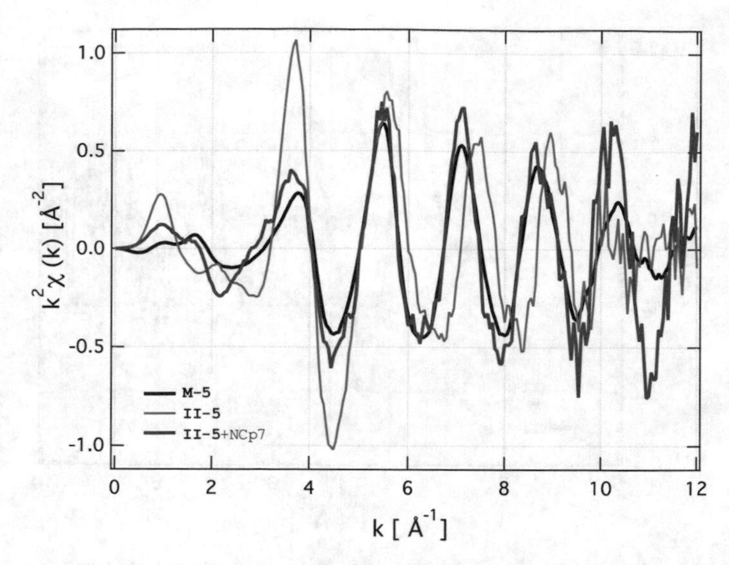

Fig. 6.8 Comparison of the k^2-weighted EXAFS of model compound [Au(*N*-Ac-Cys)] (**M-5**), with coordination sphere S-Au-S, compound $[AuCl(dien)]^{2+}$ (**II-5**) and the reaction product **II-5**+NCp7

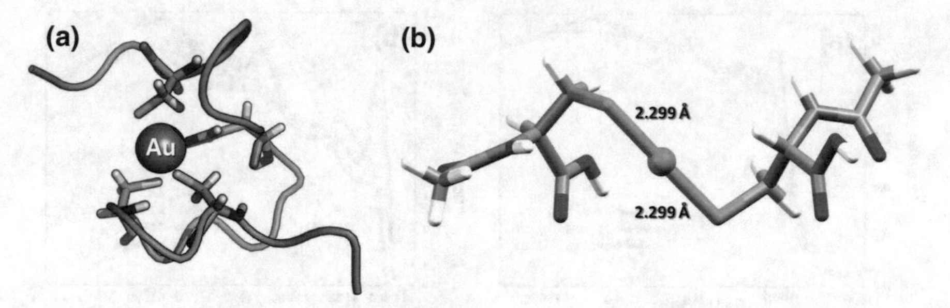

Fig. 6.9 **a** The AuF obtained when interacting **II-5**+NCp7 is expected to have a similar S-Au-S coordination as observed for the compound [Au(*N*-Ac-Cys)] (**M-5**). **b** DFT-optimized structure of $[Au(N\text{-Ac-Cys})_2]$, highlighting the Au-S distance of about 2.30 Å

EXAFS

See Figs. 6.8, 6.9, 6.10 and 6.11.

Mass Spectrometry

See Fig. 6.12.

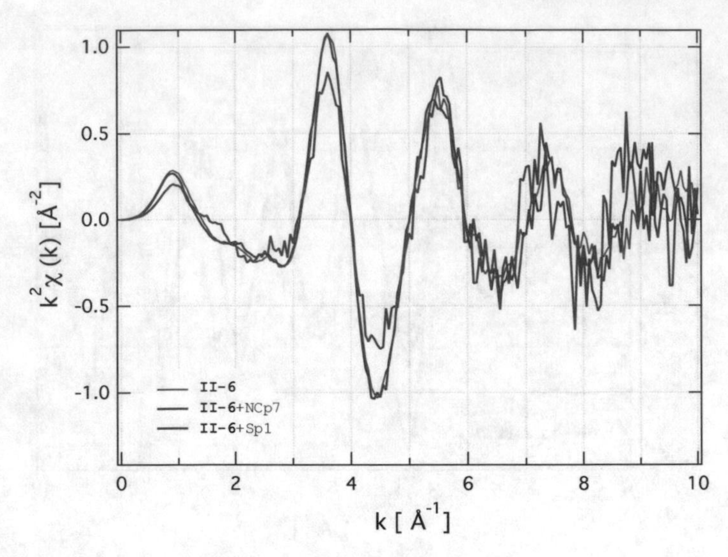

Fig. 6.10 Comparison of the k^2-weighted EXAFS of model compound $[Au(dien)(dmap)]^{3+}$ and the reaction products **II-6**+NCp7 and **II-6**+Sp1

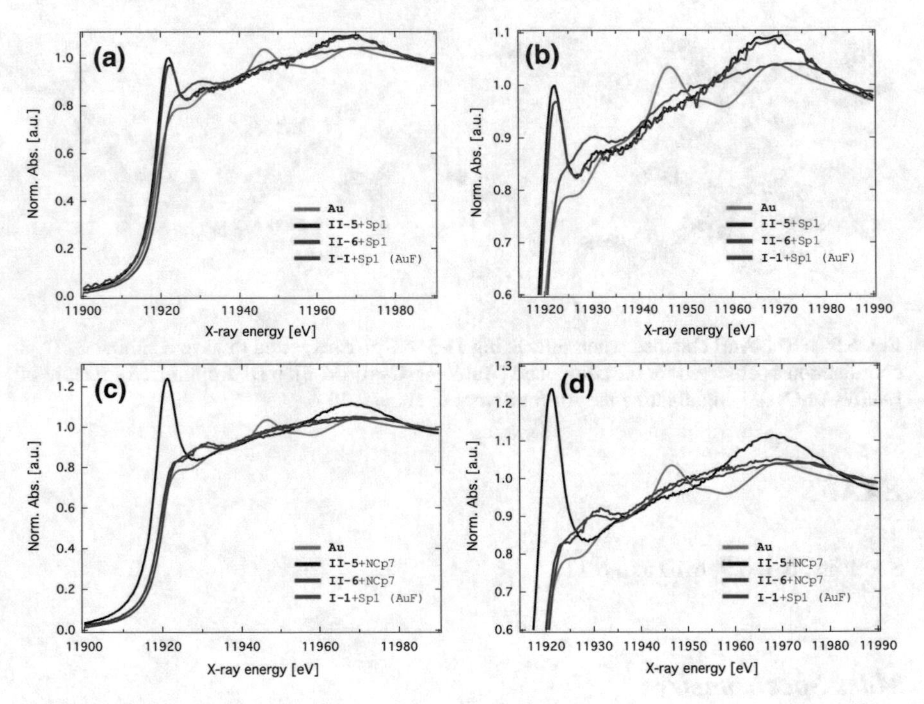

Fig. 6.11 Experimental Au L_3-edge spectra of **a II-5**+Sp1 and **II-6**+Sp1 and **c II-5**+NCp7 and **II-6**+NCp7. The near-edge regions are shown in (**b**) and (**d**)

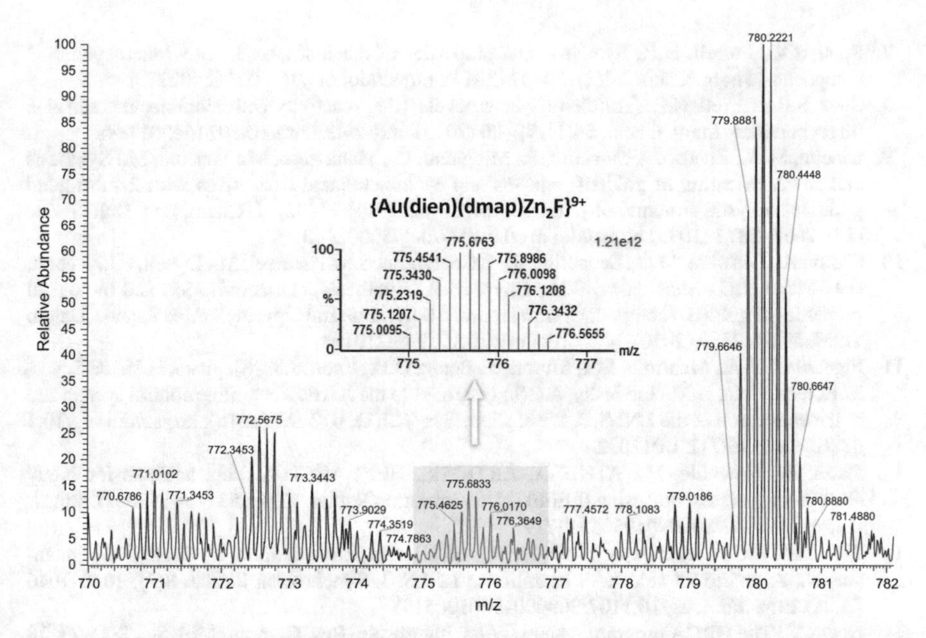

Fig. 6.12 Non-covalent adduct identified between compound **II-6** ([Au(dien)(dmap)]$^{3+}$ and the dinuclear full-length NCp7 ZnF

References

1. Berners-Price, S.J., Filipovska, A.: Gold compounds as therapeutic agents for human diseases. Metallomics **3**(9), 863 (2011). https://doi.org/10.1039/c1mt00062d
2. Milacic, V., Chen, D., Ronconi, L., Landis-Piwowar, K.R., Fregona, D., Dou, Q.P.: A Novel anticancer gold(III) dithiocarbamate compound inhibits the activity of a purified 20S proteasome and 26S proteasome in human breast cancer cell cultures and xenografts. Cancer Res. **66**(21), 10478–10486 (2006). https://doi.org/10.1158/0008-5472.CAN-06-3017
3. Ronconi, L., Giovagnini, L., Marzano, C., Bettìo, F., Graziani, R., Pilloni, G., Fregona, D.: Gold dithiocarbamate derivatives as potential antineoplastic agents: Design, spectroscopic properties, and in vitro antitumor activity. Inorg. Chem. **44**(6), 1867–1881 (2005). https://doi.org/10.1021/ic048260v
4. Che, C.-M., Sun, R.W.-Y., Yu, W.-Y., Ko, C.-B., Zhu, N., Sun, H.: Gold(III) porphyrins as a new class of anticancer drugs: cytotoxicity, DNA binding and induction of apoptosis in human cervix epitheloid cancer cells. Chem. Commun. (14), 1718–1719 (2003) https://doi.org/10.1039/b303294a
5. Jacques, A., Lebrun, C., Casini, A., Kieffer, I., Proux, O., Latour, J.-M., Sénèque, O.: Reactivity of Cys 4 zinc finger domains with gold(III) complexes: insights into the formation of "gold fingers". Inorg. Chem. **54**(8), 4104–4113 (2015). https://doi.org/10.1021/acs.inorgchem.5b00360
6. de Paula, Q. A., Liu, Q., Almaraz, E., Denny, J.A., Mangrum, J.B., Bhuvanesh, N., Darensbourg, M.Y., Farrell, N.P.: Reactions of palladium and gold complexes with zinc-thiolate chelates using electrospray mass spectrometry and X-ray diffraction: molecular identification of [Pd(bme-dach)], [Au(bme-dach)]$^{+}$ and [ZnCl(bme-dach)]2Pd. Dalton Trans. (48), 10896–10903 (2009) https://doi.org/10.1039/b917748p

7. Spell, S.R., Farrell, N.P.: Synthesis and properties of the first [Au(dien)(N-heterocycle)] 3+ compounds. Inorg. Chem. **53**(1), 30–32 (2014). https://doi.org/10.1021/ic402772j
8. Spell, S.R., Farrell, N.P.: [Au(dien)(N-heterocycle)] 3+: reactivity with biomolecules and zinc finger peptides. Inorg. Chem. **54**(1), 79–86 (2015). https://doi.org/10.1021/ic501784n
9. Cinellu, M.A., Zucca, A., Stoccoro, S., Minghetti, G., Manassero, M., Sansoni, M.: Synthesis and characterization of gold(III) adducts and cyclometallated derivatives with 2-substituted pyridines. Crystal structure of [Au{NC5H4(CMe2C6H4)-2}Cl2]. J. Chem. Soc. Dalt. Trans. (17), 2865–2872 (1995) https://doi.org/10.1039/dt9950002865
10. Casini, A., Diawara, M.C., Scopelliti, R., Zakeeruddin, S.M., Grätzel, M., Dyson, P.J., Abbott, B.J., Mayo, J.G., Shoemaker, R.H., Boyd, M.R.: Synthesis, characterisation and biological properties of gold(III) compounds with modified bipyridine and bipyridylamine ligands. Dalton Trans. **39**(9), 2239 (2010). https://doi.org/10.1039/b921019a
11. Figueroa, S.J.A., Mauricio, J.C., Murari, J., Beniz, D.B., Piton, J.R., Slepicka, H.H., de Sousa, M.F., Espíndola, A.M., Levinsky, A.P.S.: Upgrades to the XAFS2 beamline control system and to the endstation at the LNLS. J. Phys: Conf. Ser. **712**(1), 012022 (2016). https://doi.org/10.1088/1742-6596/712/1/012022
12. Ravel, B., Newville, M.: ATHENA, ARTEMIS, HEPHAESTUS: data analysis for X-ray absorption spectroscopy using IFEFFIT. J. Synchrotron Radiat. **12**(4), 537–541 (2005). https://doi.org/10.1107/S0909049505012719
13. Tolentino, H.C.N., Ramos, A.Y., Alves, M.C.M., Barrea, R.A., Tamura, E., Cezar, J.C., Watanabe, N.: A, 2.3 to 25 keV XAS beamline at LNLS. J. Synchrotron Radiat. **8**(3), 1040–1046 (2001). https://doi.org/10.1107/S0909049501005143
14. Neese, F.: The ORCA program system. Wiley Interdiscip. Rev. Comput. Mol. Sci. **2**(1), 73–78 (2012). https://doi.org/10.1002/wcms.81
15. Weigend, F., Ahlrichs, R., Peterson, K.A., Dunning, T.H., Pitzer, R.M., Bergner, A.: Balanced basis sets of split valence, triple zeta valence and quadruple zeta valence quality for H to Rn: Design and assessment of accuracy. Phys. Chem. Chem. Phys. **7**(18), 3297 (2005). https://doi.org/10.1039/b508541a
16. Petrenko, T., Kossmann, S., Neese, F.: Efficient time-dependent density functional theory approximations for hybrid density functionals: analytical gradients and parallelization. J. Chem. Phys. **134**(5), 054116 (2011). https://doi.org/10.1063/1.3533441
17. Perdew, J.P., Burke, K., Ernzerhof, M.: Generalized gradient approximation made simple. Phys. Rev. Lett. **77**(18), 3865–3868 (1996). https://doi.org/10.1103/PhysRevLett.77.3865
18. Perdew, J.P., Ernzerhof, M., Burke, K.: Rationale for mixing exact exchange with density functional approximations. J. Chem. Phys. **105**(22), 9982–9985 (1996). https://doi.org/10.1063/1.472933
19. Adamo, C., Barone, V.: Toward reliable density functional methods without adjustable parameters: the PBE0 model. J. Chem. Phys. **110**(13), 6158–6170 (1999). https://doi.org/10.1063/1.478522
20. Hess, B.A.: Applicability of the no-pair equation with free-particle projection operators to atomic and molecular structure calculations. Phys. Rev. A **32**(2), 756–763 (1985). https://doi.org/10.1103/PhysRevA.32.756
21. Hess, B.A.: Relativistic electronic-structure calculations employing a two-component no-pair formalism with external-field projection operators. Phys. Rev. A **33**(6), 3742–3748 (1986). https://doi.org/10.1103/PhysRevA.33.3742
22. Jansen, G., Hess, B.A.: Revision of the Douglas-Kroll transformation. Phys. Rev. A **39**(11), 6016–6017 (1989). https://doi.org/10.1103/PhysRevA.39.6016
23. Pantazis, D.A., Chen, X.-Y., Landis, C.R., Neese, F.: All-electron scalar relativistic basis sets for third-row transition metal atoms. J. Chem. Theory Comput. **4**(6), 908–919 (2008). https://doi.org/10.1021/ct800047t
24. Izsák, R., Neese, F.: An overlap fitted chain of spheres exchange method. J. Chem. Phys. **135**(14), 144105 (2011). https://doi.org/10.1063/1.3646921

25. Neese, F., Wennmohs, F., Hansen, A., Becker, U.: Efficient, approximate and parallel Hartree-Fock and hybrid DFT calculations. A 'chain-of-spheres' algorithm for the Hartree-Fock exchange. Chem. Phys. **356**(1), 98–109 (2009). https://doi.org/10.1016/j.chemphys.2008.10.036
26. Messori, L., Balerna, A., Ascone, I., Castellano, C., Gabbiani, C., Casini, A., Marchioni, C., Jaouen, G., Congiu Castellano, A.: X-ray absorption spectroscopy studies of the adducts formed between cytotoxic gold compounds and two major serum proteins. JBIC, J. Biol. Inorg. Chem. **16**(3), 491–499 (2011). https://doi.org/10.1007/s00775-010-0748-5
27. Gabbiani, C., Massai, L., Scaletti, F., Michelucci, E., Maiore, L., Cinellu, M.A., Messori, L.: Protein metalation by metal-based drugs: reactions of cytotoxic gold compounds with cytochrome c and lysozyme. JBIC, J. Biol. Inorg. Chem. **17**(8), 1293–1302 (2012). https://doi.org/10.1007/s00775-012-0952-6
28. Garg, D., Torbett, B.E.: Advances in targeting nucleocapsid–nucleic acid interactions in HIV-1 therapy. Virus Res. **193**, 135–143 (2014). https://doi.org/10.1016/j.virusres.2014.07.004
29. Spell, S.R., Mangrum, J.B., Peterson, E.J., Fabris, D., Ptak, R., Farrell, N.P.: Au(<scp>iii</scp>) compounds as HIV nucleocapsid protein (NCp7)–nucleic acid antagonists. Chem. Commun. **53**(1), 91–94 (2017). https://doi.org/10.1039/C6CC07970A
30. Đurović, M.D., Bugarčić, Ž.D., Heinemann, F.W., van Eldik, R.: Substitution versus redox reactions of gold(III) complexes with L-cysteine, L-methionine and glutathione. Dalton Trans. **43**(10), 3911–3921 (2014). https://doi.org/10.1039/c3dt53140f
31. Djeković, A., Petrović, B., Bugarčić, Ž.D., Puchta, R., van Eldik, R.: Kinetics and mechanism of the reactions of Au(III) complexes with some biologically relevant molecules. Dalton Trans. **41**(13), 3633–3641 (2012). https://doi.org/10.1039/c2dt11843b
32. Summers, M.F., Henderson, L.E., Chance, M.R., South, T.L., Blake, P.R., Perez-Alvarado, G., Bess, J.W., Sowder, R.C., Arthur, L.O., Sagi, I., et al.: Nucleocapsid zinc fingers detected in retroviruses: EXAFS studies of intact viruses and the solution-state structure of the nucleocapsid protein from HIV-1. Protein Sci. **1**(5), 563–574 (1992). https://doi.org/10.1002/pro.5560010502
33. Giachini, L., Veronesi, G., Francia, F., Venturoli, G., Boscherini, F.: Synergic approach to XAFS analysis for the identification of most probable binding motifs for mononuclear zinc sites in metalloproteins. J. Synchrotron Radiat. **17**(1), 41–52 (2010) https://doi.org/10.1107/s090904950904919x
34. Mijovilovich, A., Meyer-Klaucke, W.: Simulating the XANES of metalloenzymes ? A case study. J. Synchrotron Radiat. **10**(1), 64–68 (2003) https://doi.org/10.1107/s0909049502017296
35. Laskay, Ü.A., Garino, C., Tsybin, Y.O., Salassa, L., Casini, A., Laskay, U.A., Garino, C., Tsybin, Y.O., Salassa, L., Casini, A.: Gold finger formation studied by high-resolution mass spectrometry and in silico methods. Chem. Commun. **51**(9), 1612–1615 (2015). https://doi.org/10.1039/C4CC07490D
36. Chang, S.-Y., Uehara, A., Booth, S.G., Ignatyev, K., Mosselmans, J.F.W., Dryfe, R.A.W., Schroeder, S.L.M.: Structure and bonding in Au(I) chloride species: a critical examination of X-ray absorption spectroscopy (XAS) data. RSC Adv. **5**(9), 6912–6918 (2015). https://doi.org/10.1039/C4RA13087A

Printed in the United States
By Bookmasters